CONTENTS

WITNESSES

THURSDAY, DECEMBER 8, 2005

ALPHABETICAL LIST OF WITNESSES

APPENDIX

S. Hrg. 109–591

HURRICANE KATRINA: PERSPECTIVES OF FEMA'S OPERATIONS PROFESSIONALS

HEARING

BEFORE THE

COMMITTEE ON HOMELAND SECURITY AND GOVERNMENTAL AFFAIRS UNITED STATES SENATE

ONE HUNDRED NINTH CONGRESS

FIRST SESSION

DECEMBER 8, 2005

Printed for the use of the
Committee on Homeland Security and Governmental Affairs

U.S. GOVERNMENT PRINTING OFFICE
26–744 PDF WASHINGTON : 2006

For sale by the Superintendent of Documents, U.S. Government Printing Office
Internet: bookstore.gpo.gov Phone: toll free (866) 512–1800; DC area (202) 512–1800
Fax: (202) 512–2250 Mail: Stop SSOP, Washington, DC 20402–0001

COMMITTEE ON HOMELAND SECURITY AND GOVERNMENTAL AFFAIRS

SUSAN M. COLLINS, Maine, *Chairman*

TED STEVENS, Alaska
GEORGE V. VOINOVICH, Ohio
NORM COLEMAN, Minnesota
TOM COBURN, Oklahoma
LINCOLN D. CHAFEE, Rhode Island
ROBERT F. BENNETT, Utah
PETE V. DOMENICI, New Mexico
JOHN W. WARNER, Virginia

JOSEPH I. LIEBERMAN, Connecticut
CARL LEVIN, Michigan
DANIEL K. AKAKA, Hawaii
THOMAS R. CARPER, Delaware
MARK DAYTON, Minnesota
FRANK LAUTENBERG, New Jersey
MARK PRYOR, Arkansas

MICHAEL D. BOPP, *Staff Director and Chief Counsel*
DAVID T. FLANAGAN, *General Counsel*
JONATHAN T. NASS, *Counsel*
JOYCE A. RECHTSCHAFFEN, *Minority Staff Director and Counsel*
ROBERT F. MUSE, *Minority General Counsel*
MARY BETH SCHULTZ, *Counsel*
TRINA DRIESSNACK TYRER, *Chief Clerk*

HURRICANE KATRINA: PERSPECTIVES OF FEMA'S OPERATIONS PROFESSIONALS

THURSDAY, DECEMBER 8, 2005

U.S. SENATE,
COMMITTEE ON HOMELAND SECURITY
AND GOVERNMENTAL AFFAIRS,
Washington, DC.

The Committee met, pursuant to notice, at 10:02 a.m., in room SD–342, Dirksen Senate Office Building, the Hon. Susan M. Collins, Chairman of the Committee, presiding.

Present: Senators Collins, Stevens, and Lieberman.

OPENING STATEMENT OF CHAIRMAN COLLINS

Chairman COLLINS. The Committee will come to order. Today, the Committee continues its investigation into the preparation for and response to Hurricane Katrina.

This morning, at our eighth hearing, we will hear from three witnesses who are among the Federal Emergency Management Agency's most experienced emergency managers and operations professionals. Each of our witnesses—Scott Wells, Philip Parr, and William Carwile—was directly involved in Katrina preparation and response. They will give us a more complete understanding of FEMA's role, share their observations about the State and local response, and provide their insights and recommendations for reforms.

One of our witnesses today has described the national emergency response as a "bottom-up system," with local and State authorities leading the way and Federal authorities coordinating operations and the deployment of resources. In Katrina, this system broke down, and the result was the very deprivation and suffering the structure was designed to avoid. This system must be fixed from the bottom to the very top.

One of the most glaring breakdowns was in communications. This powerful storm devastated the land-based communications infrastructure throughout the Gulf region. This, however, was an utterly foreseeable result of howling winds and surging water that apparently was not adequately anticipated, nor compensated for.

At our last hearing on November 16, we heard testimony from private sector witnesses who stressed the critical importance of maintaining communications in disaster management. They emphasized that good communications are the life blood of emergency operations, allowing for the effective movement of personnel and other assets as well as real-time assessments.

In each of these companies, developing and maintaining robust systems, importing extra communications gear, and re-establishing contact with the outside world were of the utmost priority and a key component of their preparedness plans. Their outstanding performance, unfortunately, stands in stark contrast to the inability of government at all levels to plan and execute backup communications systems.

FEMA has mobile communications vehicles. But by the time anyone thought to bring one to the Superdome, the building was already surrounded by water, and FEMA was apparently unable to figure out a way to get its equipment into the building.

FEMA also has communications equipment that could be airlifted in. But despite Mr. Parr's urgent request for such equipment, none arrived. In his interview with the staff, Mr. Parr estimated that the lack of communications equipment reduced his team's effectiveness by an astounding 90 percent.

Much of the post-Katrina criticism has been justifiably focused on FEMA. But today's witnesses will explain that Katrina also exposed serious flaws at the local and State level that contributed to the suffering experienced by so many in the Gulf region.

For example, according to the staff interview of Mr. Wells, Louisiana's emergency operations officials failed to follow—perhaps even to comprehend—the National Response Plan, which is an integrated system designed to coordinate the Federal, State, and local responses to a disaster. Indeed, Mr. Wells noted that Louisiana's emergency managers were getting training on the critical Incident Command System 2 days after the storm hit.

Today's witnesses will also help us determine how FEMA, State, and local officials can do better. They are all current or former Federal Coordinating Officers (FCOs) and possess a wealth of emergency management experience. The FCOs play a critical role in FEMA.

In June of 2004, the FCO cadre urged Under Secretary Michael Brown to undertake reforms to remove obstacles to command, control, and core mission accomplishment and to reconfigure and enhance the national emergency response teams. The memorandum strongly advised that these reforms be implemented to help prepare for "the next big one."

But we will hear today that disturbingly little was done in response to these recommendations, far too little to prepare for the big one when, indeed, it hit 14 months later.

I very much appreciate the testimony of our witnesses today so that we can be better prepared for the catastrophic events yet to come.

Senator Lieberman.

OPENING STATEMENT OF SENATOR LIEBERMAN

Senator LIEBERMAN. Thanks very much, Madam Chairman, for calling this hearing and for your continued leadership of our investigation.

We have already held several important and informative hearings about Hurricane Katrina on subjects that range from the fate of the levees to the actions of the private sector in the wake of the storm, to the testimony of FEMA employee Marty Bahamonde during those dangerous days when he tried to get help to the thousands stranded in the Superdome.

This morning's witnesses, FEMA's operations professionals who were on the scene in the Gulf Coast, will, I am confident, move the Committee's investigation forward toward finding out what went wrong before and after Hurricane Katrina so that we can achieve the purpose that Congress and all of us have in mind, which is to make sure that it never happens again.

I want to thank our witnesses for the testimony they are about to give. I want to thank them for their distinguished careers of public service that each of them has. And those careers should not, in any sense, be diminished by the criticism that FEMA is receiving and may well receive today.

I will say that, having reviewed your testimony prepared for this morning and the interviews that you had with our Committee staff and having now gone over other testimony and documents gathered by the Committee so far, it certainly seems to me that FEMA is a troubled agency that failed at its prime mission, the mission it draws its name from, which is emergency management.

The fact is that the whole world watched on television as Hurricane Katrina, a disaster waiting to happen, developed in the Gulf of Mexico. The whole world listened to the experts who said that this was the long-dreaded "big one" that could take out the levees and flood the Big Easy. Yet FEMA seemed to underestimate the gravity of the storm coming and/or failed to realize that doing business as usual was unacceptable and would compound the disaster.

Katrina obviously was not a typical hurricane in response to which FEMA or anyone else—Federal, State, or local—could work off of a typical playbook. Katrina required a more urgent, comprehensive, and aggressive governmental response.

Katrina was a catastrophe. It knocked out many of the State and local communications, as Senator Collins has said, and response capabilities and overwhelmed those that remained. But FEMA seemed to expect a severely damaged State and local response network, itself the victim of the catastrophe, to operate as if it was at full and normal capacity.

Like Senator Collins, I have been very surprised and upset to learn in the course of our Committee's investigation that America's battle plan for catastrophes, the National Response Plan's Catastrophic Incident Annex, was never activated in response to Katrina. And FEMA apparently still believes that it should not have been activated.

As we will hear today, FEMA deployed too few people to respond to Katrina and deployed them too slowly. Many of those that did deploy apparently failed to appreciate what the breaking of the levees around New Orleans meant, and that failure had disastrous consequences, as we all know, for the people of New Orleans.

As we learned at our previous hearings, New Orleans industrial canal levees were leveled by the storm surge early Monday morning, August 29. That led to almost immediate flooding in the eastern part of the city, including the lower 9th Ward. By mid day, the Lake Pontchartrain levees were also breaking, and that led to a much slower flooding of downtown New Orleans, what we so often heard referred to as New Orleans "filling up like a bowl."

Mr. Bahamonde, previously referred to as FEMA's first man on the ground in New Orleans, told us that he communicated these facts by mid day Monday to FEMA and had a conference call with FEMA officials at the emergency operations center, among others, that night. We now know that other sources were providing the same information throughout the day to the Baton Rouge emergency operations center, where FEMA's top regional operators were stationed. Yet, as we will hear today, the FEMA emergency response team did not depart Baton Rouge for New Orleans until noon on Tuesday, significant hours later, almost a full day after the hurricane had hit and already passed.

By that time, Lake Pontchartrain had been dumping its waters into downtown New Orleans for hours, making it impossible for the FEMA team to bring its vital communications tractor-trailer, so-called "Red October," into the city. This left the team without any reliable means of communications and reduced its effectiveness in New Orleans, as Senator Collins said, by some 90 percent. That is according to Mr. Parr's testimony that we will hear this morning.

But that wasn't the only costly delay. Unfortunately, we have learned from other witnesses that the Coast Guard was performing rescue missions as soon as hurricane-force winds abated on Monday afternoon. The State itself sent out rescue boats later Monday afternoon. But FEMA's search and rescue teams didn't arrive in New Orleans until Tuesday morning, and we want to ask why.

Given the catastrophic nature of Katrina's damage, we must understand why FEMA wasn't prepared to move sooner. And of course, the most vexing part of it all is not just that this was all foreseeable, but that, in fact, it had been foreseen. This precise disaster scenario was used in the Hurricane Pam planning exercise conducted in June 2004. It also had been the topic of numerous stories in the media and hurricane conferences over the years.

This was not a failure of imagination, as some might want to label it. It was a failure of realization. Realization that the catastrophe, about which we had all long been warned, was about to occur and that FEMA and everybody else, State and local, had to move quickly to address it.

Yes, a disaster like Hurricane Katrina is an act of God. Yes, there will be confusion in such extraordinary natural disasters. Yes, mistakes will be made by people who are well intentioned. But adequately preparing for and responding to a disaster of this magnitude required a well-led, well-trained, well-drilled, and well-manned FEMA that had a plan in place and a sense of mission to guide its actions.

Regrettably, it appears to me, at this point in our investigation, that all of these things were lacking as disaster swept across the Gulf Coast region last August. This morning, we want to ask why. And I am confident that these three witnesses can help us answer that question.

Thank you very much, Madam Chairman.

Chairman COLLINS. Thank you. Senator Stevens.

OPENING STATEMENT OF SENATOR STEVENS

Senator STEVENS. Thank you, Madam Chairman and Senator Lieberman.

I regret to disagree with the Senator from Connecticut. Our Committee took a trip to New Orleans. As a result of that trip, we became convinced that what happened in New Orleans was that Katrina went through New Orleans. We saw the buses in line that were ready to deal with the evacuation. The people of New Orleans were prepared for Hurricane Katrina. They were not prepared for the failure of the man-made systems of levees and gates and the enormous impact of that canal from New Orleans to the sea, which should never have been there.

It is like saying that in terms of our earthquake, which we had in the Anchorage area, that someone was at fault because they didn't notify Kodiak and Seward that a tsunami might hit them—which did happen. What happened here is in the aftermath of Katrina going through, because of the subsequent series of events that caused the failure of the levees, the failure of the system, it was impossible to execute the plan.

Now the plan for New Orleans was caused by a provision in the 1998 appropriations bill, which the last administration failed to make. But finally, in 1999, we mandated that plan. It was prepared. It was actually exercised after the start of 2001. It was there, and I think the people of New Orleans started to follow that plan, and they started to move their people to the dome and to the various places which should have been safe.

But with the failure of the man-made systems and the failure of having the ability to shut off the surge that came across Lake Pontchartrain, this became a man-made disaster. And I do not agree that we can fault FEMA or the City of New Orleans or the State of Louisiana for failing to anticipate the complete failure of the systems that were prepared in the past.

As a matter of fact, I think you can go back to President Johnson's time and find that he tried to build even better systems at the time, and the funding was turned down.

But as a practical matter, this damage, as sad as it is—it is a sad thing—it is not a failure of the warning system. The warning was there. It was a failure of systems that were put in over the last 30 years to prevent the surges that happened in a way and the combination of them was by the time that surge came in from the ocean, it came up that canal, it was like a tsunami coming up that canal. Had the canal not been there, those levees might not have failed.

So this is not some time to critique the failure of the people involved to predict that the basic systems for protection that had been designed over a period of years would simultaneously all collapse. And that is what happened. Every single one of them collapsed.

Now those were man-made. This isn't an act of God. This is a failure of our basic engineering systems, our basic concepts of protection, and we have to do better in the future. I am not going to join in criticizing those who tried to do the best in the most extraordinary circumstance I have ever seen.

Now this Senator has seen war. I have seen cities in China totally destroyed. I never saw destruction like I saw in New Orleans. No person on Earth could have predicted that. It had to be a combination of circumstances caused by Katrina going through, and then the surge and all the collapse that came afterward.

So I hope we look at FEMA and the rest of these people and ask them what can we do to prevent this in the future? Let us quit looking backwards and trying to assess blame. Let us find out what can we do to assure that this won't happen not only in New Orleans, but anywhere else where we are relying on levees and manmade protections to prevent disasters.

Thank you very much.

Chairman COLLINS. I would now like to welcome our panel of witnesses before us today.

Our first witness, Scott Wells, joined FEMA as a Federal Coordinating Officer in 1999. Since that time, he has been deployed by FEMA to more than 20 disasters. For Hurricane Katrina, he was the second in command, serving as the deputy FCO.

Mr. Wells arrived at the emergency operations center in Baton Rouge on Saturday, August 27. Mr. Wells previously served as an Army officer for more than 20 years, and he was the DOD liaison to FEMA before retiring from the Army in 1999.

Next we will hear from Philip Parr. For more than 20 years, Mr. Parr served as a firefighter for New York City, ultimately rising to the rank of battalion chief. In January 2004, Mr. Parr left the New York City Fire Department and joined FEMA as a Federal Coordinating Officer.

During Hurricane Katrina, Mr. Parr led the emergency response advance team and was deployed to New Orleans. Mr. Parr and members of his team arrived in the Superdome on the morning of Tuesday, August 30. He later led FEMA teams into the hardest-hit parishes of southern Louisiana.

William Carwile joined FEMA in 1996 as director of Region 10, headquartered in the Pacific, and was a Federal Coordinating Officer for five tropical storms in the Pacific. His emergency management experiences on the mainland include New York City following September 11, the 2003 California wildfires, and four hurricanes that struck Florida last year.

During Hurricane Katrina, he served as the Federal Coordinating Officer for Mississippi. He is a retired U.S. Army colonel with a 30-year military career.

I want to thank each of you for your testimony today and for your public service. Because this is part of an ongoing investigation, I would ask that you each stand so that I can swear you in. Please raise your right hand.

Do you swear that the testimony you are about to give to the Committee will be the truth, the whole truth, and nothing but the truth, so help you, God?

The WITNESSES. I do.

Chairman COLLINS. Thank you.

Mr. Wells, if you would go first and proceed with your statement?

TESTIMONY OF SCOTT WELLS,[1] FEDERAL COORDINATING OFFICER, FEMA JOINT FIELD OFFICE, BATON ROUGE, LOUISIANA

Mr. WELLS. Good morning, Chairman Collins and Members of the Committee.

My name is Scott Wells, and I am honored to appear before you today. My current position with FEMA is Federal Coordinating Officer in Louisiana for Hurricanes Katrina and Rita.

I would like to start this morning by thanking you for the invitation to testify before this Committee. I appreciate the opportunity to share my perspectives on FEMA operations before, during, and after our Nation's costliest disaster, Hurricane Katrina.

It is my intention today to speak candidly with you about my experiences in Louisiana, both leading up to and following Katrina, as well as my perspectives on emergency management. I will begin my testimony today with a brief overview of my professional career in emergency management.

For almost 2 decades, I served in various positions of emergency management. Beginning in 1985, for 2 years as a first responder MEDEVAC pilot. During my 24-year military career, I also spent 10 years in the Pentagon providing military support to civilian authorities. My last military assignment in the Pentagon was as a military liaison officer to FEMA.

In these assignments, I was involved in numerous disasters and emergencies such as Hurricane Andrew, the Northridge earthquake, the Midwest floods of 1993, the Oklahoma City bombing, the Haitian/Cuban immigration emergency, the Waco siege, and the Ruby Ridge incident.

I retired in 1999 from the Army and have been working for the Federal Emergency Management Agency as an FCO for the past 6 years and served on several disasters and emergencies to include Tropical Storm Allison in Texas and the Columbia Shuttle recovery operation.

On August 27, I was assigned to Louisiana as the Deputy Federal Coordinating Officer to Bill Lokey for Hurricane Katrina. I served in that capacity until September 19. At that time, I was reassigned to Texas as the Federal Coordinating Officer for Hurricane Rita, as she made her way through the Gulf of Mexico. As the FCO for Rita, I remained in Texas until the first week of October, at which time I returned to Louisiana to replace Bill Lokey for Hurricane Katrina.

Detailed preparation for a Katrina landfall in Louisiana started in earnest on Friday, August 26, when the National Weather Service quickly changed the projected zone for landfall to include Louisiana. Much work had been done earlier in the week, but the focus of those efforts—given the projected path of the storm—was on Florida, Alabama, and Mississippi.

Field deployments to Louisiana began on Saturday, August 27, with the emergency response team's advance elements deploying to the Louisiana emergency operations center here in Baton Rouge. The Federal regional and national staffs consolidated that night and started conducting field operations in preparation for landfall.

[1] The prepared statement of Mr. Wells appears in the Appendix on page 43.

In addition to the command and control element being set up in Baton Rouge, we concurrently were setting up an operational staging area in Alexandria, Louisiana, that served as a Federal logistics base for Katrina operations.

The first actual employment of Federal resources—that is, where we provided—FEMA provided response assets to the State—occurred on Sunday, the day before landfall, when we shipped six truckloads of water and three truckloads of MREs to the Superdome. Two of the truckloads, one each of water and MRE, did not complete delivery. They were denied entry by the Louisiana State Police before reaching the Superdome.

That was the beginning of response operations and was soon followed by many other response resources, such as medical teams, search and rescue teams, and other critical commodities, such as water, food, and ice.

There has been much said about the slow Federal response to Katrina. From my perspective, with all due respect to Senator Lieberman, I want to say nothing could be further from the truth. We had a fully operational logistics base, a fully operational command cell, and response teams in place, all before landfall. We even moved some supplies in before landfall and attempted to move in a medical team.

On the day of landfall, we moved search and rescue teams, medical teams, and critical supplies into the affected area. It may not have been enough for an event of this magnitude, but it was fast.

I think the real issue is that the response was not robust. It was not enough for the catastrophe at hand. And as you look—as we all look—to make it better next time, I think it is an important distinction to make. "Slow" means one thing. "Not enough" means something else. More importantly, the corrective actions between fixing "slow" and "not enough" could be significant.

Emergency management is unlike any other system in the government. It is a bottoms-up approach. The people on the ground are in charge. The first responders are supported, as required, by local government, then State government, and as a last resort, the Federal Government. Ultimately, authority for disaster response operations rests at the local level. The State and Federal Governments are not in charge, but are responsible for assisting local governments.

And that is how it should be, as all disasters are local. Disasters start at the local level, and disasters end at the local level. This system works for small to medium disasters. It does not work so well for large disasters, and it falls apart for a catastrophic disaster. I think that is a fundamental problem with the response to Katrina.

Following are some of the other major problems and proposed changes I believe could improve our national readiness posture to respond to future disasters. And I will list them, but in the interest of time, I won't go through all of my statement, but you will have it.

(1) We need to strengthen the emergency management capability at the State and local level.

(2) We need to review the emergency management architecture for response and recovery operations. There are problems associ-

ated with the implementation of the Stafford Act as it is executed through the National Response Plan and the Incident Command System.

(3) We need a trained, staffed, and equipped Federal response team.

(4) We need to change the financial management of disasters.

(5) We need to simplify the public assistance process.

(6) We need to simplify individual assistance process.

(7) We need a greater investment in the leadership and management within FEMA.

Emergency management is not a simple system. Accordingly, there are no simple solutions. To have an effective national disaster response structure, we must have a viable local, State, and Federal capability. If any of these links in the emergency management chain breaks, the system itself begins to break down. If we cannot have viability at all three levels, then we should change the system.

Thank you for giving me the opportunity to share a field perspective of Hurricane Katrina.

Chairman COLLINS. Thank you, Mr. Wells. Mr. Parr.

TESTIMONY OF PHILIP E. PARR,[1] DEPUTY FEDERAL COORDINATING OFFICER, FEMA JOINT FIELD OFFICE, AUSTIN, TEXAS

Mr. PARR. First, I want to say good morning to this august Committee. Good morning, Chairman Collins.

My name is Phil Parr, and I want to thank you for the opportunity of testifying before you about my experiences and the response to Hurricane Katrina. The views expressed in my testimony are my own and do not necessarily reflect the views of the Department of Homeland Security.

Presently, my position with FEMA is that of Federal Coordinating Officer. I have been involved with response and emergency management for the past 26 years. I was sworn in as a member of the New York City Fire Department in 1979 and rose through the ranks to attain the level of chief officer in 1999.

During my tenure with the FDNY, and particularly during my tenure as a chief officer, I served in many capacities, including, but not limited to, fire and emergency ground commander, operations, planning for Y2K scenarios, and as a deputy director in the New York City Office of Emergency Management. I have played an active role in countless disasters and crisis situations to include the September 11 attack at the World Trade Center, where I was on scene prior to the towers collapse.

Since January 2004, I have been a member of the Federal Coordinating Officer program, assigned to FEMA Region 1, New England. In this role, I have served in various capacities, including FCO for three presidentially declared disasters.

Before I continue with my testimony, I think it is important to mention that I have always taken great pride in my years of service as a member of the New York City Fire Department. At one

[1] The prepared statement of Mr. Parr appears in the Appendix on page 52.

time, I would not imagine serving in any other position in which I would serve with the same feeling and pride.

However, during my tenure with FEMA, the dedication to service as displayed by its members and their care for disaster victims has allowed me to serve with the same pride and satisfaction that I experienced during my previous 25 years of public service. So it is with that passion that I speak before this Committee, and I thank you again for the opportunity to do so.

On Saturday, August 27, I was informed that I would be the emergency response team advance element team leader for the State of Texas. My team was composed of personnel from FEMA Region 1, New England, and we were instructed to rendezvous in the Region 6 Regional Response Coordination Center in Denton, Texas, on Sunday, August 28. Soon it became clear that Texas was not in the path of Hurricane Katrina and that members of my team and I would be assigned as the lead element in New Orleans, Louisiana.

I flew to Louisiana immediately following the hurricane passing, Monday, August 29, with a contingent of my team. And Tuesday morning, August 30, we helicoptered into the Superdome. Our mission was threefold. One, form a unified command with the State, as represented by the Louisiana National Guard, and the City of New Orleans. Two, maintain visibility of commodities ordered. And three, build out a base from which FEMA teams could be formed to locate and assist in the hardest-hit parishes.

To accomplish these goals, we were to meet a mobile emergency operations and communications vehicle and use that as a base of operations and communication. Due to extensive flooding in the city, our communications vehicle was unable to enter the Superdome, and this severely hampered our operations.

Despite this, and while working under the most difficult of circumstances, we were able to assist the National Guard in maintaining a supply of food and water to Superdome evacuees—all were fed and provided water—and, even with limited communications, facilitate the arrival of what was to become over the next 4 days a thousand-bus convoy to evacuate the City of New Orleans to start the day after our arrival.

The FEMA disaster medical assistance team treated hundreds and identified seriously injured and special needs patients who were evacuated via air and ground assets throughout the operation. In addition, several meetings were held with the mayor and his staff, ranking National Guard officers on the scene, and other Federal officials to include DOD and the Coast Guard. This facilitated the initiation of a unified command structure.

Due to the enormity of the event, not all of our initial goals were met. A delay ensued in placing teams into other hard-hit parishes, which I believe took place that Friday and Saturday.

I have been asked whether FEMA was overwhelmed, or could our response be considered slow? To consider the latter first, I must say, in my opinion, no. FEMA teams—response, management, medical, and urban search and rescue—were in position in four States pre-landfall. Commodities were staged close to the impacted areas, and in some cases, the hand-off to the State had already taken place.

In addition, and as previously mentioned, FEMA mission assigned Emergency Support Function 1, the Department of Transportation, and they verified that, by September 3, 990 buses were in service performing evacuations. It is estimated that over 66,000 persons were transported by that date. The number of buses grew to over 1,100 in the next 2 days.

Were we overwhelmed? The simple answer is yes. But what needs to be understood is that, at any disaster, the initial response always feels overwhelmed. I must draw on my experience as a local responder to give you an example on a small scale of what I mean, and then a larger one.

The police officer who pulls up to a 2-car accident with severe injuries while he operates alone, waiting for help, is overwhelmed. The fire officer who pulls up to a burning structure with people trapped inside is overwhelmed. But the true professional, while responding and operating, knows that he is constantly sizing up the situation, gaining intelligence, shifting strategies, modifying plans, and calling for assistance where needed to meet unfulfilled needs, whether expected or unexpected.

I would like to refer back to the disaster of September 11 and its effect on the emergency personnel operating at the World Trade Center. First, it must be remembered that within the 369 square miles of New York City are the resources of a State with a strong central government. There are over 35,000 New York City police officers, about 13,000 firefighters and emergency medical personnel. These numbers only begin to enumerate the assets available to the city. No other city in the country can begin to come close to the responders that are contained within New York City.

The response to the attacks on the towers was immediate. The enormity of the task at hand was overwhelming. Then with the collapse of the towers, it was chaos. Emergency services within New York regrouped almost immediately and restarted operations, but a full, coordinated plan took days.

The World Trade Center complex was 13 acres. The landfall of Hurricane Katrina affected four States and covered an area of 90,000 square miles, an area the size of Great Britain. It affected millions.

Effectively, Louisiana was hit by two disasters. First, a devastating hurricane along with its associated blast damage and, second, a catastrophic flooding event caused by levee failures. Hurricane Katrina was the most devastating disaster to hit our country. We were all overwhelmed—the city, the State, the affected parishes, and the Federal Government.

What can FEMA, individuals, local governments, and States do to be more prepared? First, it must be realized that the response to any crisis or disaster is the responsibility of every individual and form of government in this country. Emergency management is more than just coordination. It is about partnership with all entities previously mentioned.

Each of us plays a vital part, and any one of us who fails in our part fails in that partnership. That failed responsibility must be picked up by one of their partners, and that causes delay, confusion, and lack of coordination.

For FEMA's part, it is my belief we have not done what is needed to get that message across to individuals, locals, and States. We have worked to create an image that Uncle Sam will be on your doorstep with MREs, food, water, and ice before the hurricane-force winds subside. We have created an expectation that in a large or no-notice event, such as a terrorist attack or an earthquake, we can never hope to meet.

As an agency, we must help our partners understand their role in the emergency management cycle, as many States and locals do now. To this end, I believe we can do much with conditional and competitive grants to State and local governments to achieve this.

Generally because response is immediate and local, FEMA's primary role in disaster is recovery. With some notable exceptions, what is described at the Federal level as response in actuality is "response support"—that is, supplying life-saving commodities—with local and State responders performing what we traditionally call response.

But as an agency, we can do better in the response role. Primarily, I believe this can be accomplished by a shift in attitude and training by some in management and decisionmaking roles in our agency.

In another area of improvement, FEMA has initiated a total asset visibility system whereby truckloads of commodities can be located via satellite transponder and tracked more closely. This system must be fully put online before our next hurricane season.

We should recognize that FEMA is a small agency, especially when compared with other Federal agencies. But its strength lies in the fact that the National Response Plan identifies it as the coordinating agency for the entire Federal response.

I believe more drills, familiarization and otherwise, are necessary between FEMA and other Federal agencies to help clarify roles and responsibilities under the NRP and in their critical emergency support functions. Understanding their contribution and role in the emergency response team structure is essential for effective response. These crucial elements must be established and become routine to help ensure that a better-coordinated Federal package can be delivered to States to assist them in their response.

Additional standardized and practical training must be provided to personnel who may be asked to serve on response teams at the county or local level. Training programs and expectations that build on practical experience from this and previous operations, with input from States, must be provided to FEMA staff, who may be needed to assist at the local level in response operations.

As with any operation, I hope that, as an agency, we can make these changes based on lessons learned. I would also hope that State and local officials will review their emergency management procedures and also adopt necessary changes. Finally, each citizen has a responsibility to plan, heed warnings, and do whatever is within their means to prepare and respond to disaster.

Thank you for the opportunity to speak on this subject.

Chairman COLLINS. Thank you. Mr. Carwile.

TESTIMONY OF WILLIAM L. CARWILE III,[1] FORMER FEDERAL COORDINATING OFFICER, FEMA JOINT FIELD OFFICE, BILOXI, MISSISSIPPI

Mr. CARWILE. Good morning, Chairman Collins and distinguished Members of the Committee.

I am Bill Carwile. Thank you for inviting me today to testify about operations in Mississippi during Hurricane Katrina.

Between August 29 and October 15, I was initially the Federal Coordinating Officer and later the Deputy Federal Coordinating Officer. I recently retired from FEMA and am currently affiliated with the Center for Homeland Defense and Security at the Naval Postgraduate School in Monterey, California. I am testifying today as a private citizen.

There are many lessons to be learned from the responses to Hurricane Katrina. I applaud your efforts to gather information critical to charting a future for disaster operations in our country. I hope my testimony will, in some way, make a contribution to that undertaking.

My perspective is from one who has been in the field, on the ground in large-scale disasters during much of the last 9 years. My recent experiences include serving as operations section chief for the World Trade Center operations in New York, serving as Federal Coordinating Officer for the 2003 wildfires in California, and FCO for each of the four hurricanes last year in Florida.

I joined FEMA in 1996 as the director of the Pacific area office in Honolulu, Hawaii, where I reside today. In 1999, I became one of the first members of the Federal Coordinating Officer program created to provide a pool of trained professionals to manage the Federal side of disasters.

In 2003, I was appointed as one of the first predesignated principal Federal officials by former Homeland Security Secretary Tom Ridge. Prior to my service with FEMA, I retired as an Army colonel, having served almost 3 decades as a special forces and infantry officer in the regular Army. My assignments included two tours in Vietnam.

I would like to address three major points in my oral testimony. First, there were three separate presidential disaster declarations as Hurricane Katrina struck the Gulf Coast—Louisiana, Mississippi, and Alabama. These were three different disasters in the type and extent of destruction, the number of people affected, and each State's constitutional relationships with its local jurisdictions.

While each disaster was different, they were similar in that in each disaster the governor of the State was in charge. As FCO, I was appointed by the President as his representative to support the governor using the authorities provided for in the Stafford Act. While there were three distinct disasters, today my comments will cover only operations in Mississippi, where I was.

Second, in my view, this was the first time we fully implemented appropriate portions of the National Incident Management System, the National Response Plan, and the Incident Command System in a major disaster response. During the summer of 2005, following the distribution of the National Response Plan, many Federal and

---

[1] The prepared statement of Mr. Carwile appears in the Appendix on page 61.

State emergency managers underwent training on the plan and ICS.

Fortunately, I and key members of our emergency response team in Mississippi, which is mostly comprised of personnel with whom I have worked for years, had participated in extensive ICS training. Similarly, Mississippi Emergency Management Agency Director Robert Latham and his staff and most county emergency managers had recently undergone NIMS and ICS training.

One of the key members of my team, operations section chief Bob Fenton, has long been involved in the doctrine on training development, is an expert on how to adapt ICS for large-scale operations.

Using this training and our experience in prior disasters, Robert Latham and I and our teams established a unified command to a degree beyond which I believe is envisioned by the National Response Plan and began the joint incident action planning process, which set our priorities for each of the operational periods, which is a 24-hour period at first, following the ICS concepts. Governor Barbour attended and participated in many of our meetings and provided leadership and important strategic guidance.

During the response, we found that some aspects of the National Response Plan did not fit our organizational needs for a joint State/Federal response to a catastrophic disaster. We found it necessary to modify some important aspects of the plan. These changes are detailed in my written testimony.

While it is my belief that ICS works well for fires and smaller disasters, some substantial modifications are required for large-scale events. Mostly, these modifications revolve around the need for unified command up and down the organization and in order to address political and operational realities.

I would recommend that an effort be made to capture the experiences of the individual geographic and branch directors and division supervisors. They were down in the communities and provided us a manageable span of control. This might be accomplished through interviews, similar to the Army's oral history program, to find out what really worked down there, where the rubber meets the road.

This effort will provide a more detailed view of what changes to the NRP should be made to accommodate the realities of joint State/Federal response to a catastrophic disaster.

Third, there have been questions raised about the competence of FEMA personnel in the wake of Hurricane Katrina. In my career in FEMA, I worked with many dedicated and highly competent individuals who were committed to serving both our country and the victims of disasters. Many routinely give up holidays, anniversaries, birthdays, and normal family lives to help others. I am proud to have been part of FEMA and especially proud of the many individuals whose heroic efforts helped the people of Mississippi and other States in which I have served. They deserve our thanks.

A disaster can bring out the best in people. There are thousands of stories of individual acts of heroism and kindnesses during Katrina. Mississippians helped their neighbors. Hundreds of local officials, who had just lost everything, reported for duty. And all around the country, volunteers left their lives behind and headed for the Gulf Coast to help.

We should not forget, however, that in a catastrophic disaster, the government and those wonderful voluntary agencies can never provide adequate aid in the immediate aftermath of the disaster. We all need to better prepare ourselves and our families and be ready to help our neighbors.

In my written testimony, I provide comments about what I think went well and what didn't go well in the response and initial recovery for Katrina in Mississippi. Two main points. The State/Federal unified command worked well in Mississippi. But this success was obscured by the fact that requested resources did not arrive quickly enough. Better, more effective methods must be adopted to quickly deliver resources in a catastrophic event.

In Mississippi, while temporary housing has been provided in numbers far exceeding any previous effort, this success is obscured by the overwhelming need and an exceptionally long period of time that people remain in shelters. New methodologies must be examined and implemented to take care of Americans in need of humane housing while in a catastrophic event.

These are but two of the many challenges the Nation faces if it is to really prepare for the next catastrophic disaster. We must do all we can to capture the lessons learned, both good and bad, from Katrina in Mississippi in order to make real changes so that the next time, elected and appointed officials will be able to better support the needs of victims.

I thank the Committee for undertaking this important work for the Nation. I will be glad to try to answer any questions you may have.

Chairman COLLINS. Thank you for your testimony.

I want to start my questioning today to get your judgment, your assessment of preparedness and response at the State and local level. As Mr. Wells pointed out in his testimony, our Nation's emergency management structure is often referred to as a bottom-up structure. It has key roles for local, State, and Federal Governments to play.

I thought that Mr. Wells made an important point when he said that there is some misconception on the part of the public about who plays what role in our emergency response system. So starting with you, Mr. Wells, if FEMA and the Federal Government are at the top of the structure, how did the bottom part work, in your judgment, for Hurricane Katrina? How would you assess the State and local response in Louisiana?

Mr. WELLS. Well, a lot of it is situational. Each parish in Louisiana has different capability. The important thing in emergency management is you have to have a foundation established for the Federal departments to build on.

When we go into a disaster, the locals are in charge. The locals ask for assistance from the State, who, in turn, asks for assistance from FEMA. If there is no structure, if there is no organization, if there is no capability at the local or the State level, there is no foundation from which we can build.

I will give you one example to compare between Katrina in Louisiana and Rita in Texas to try to demonstrate what I mean by that. In Texas, for Rita, I was there for the preparation phase, and we are just going to talk about prior to landfall because it is equivalent to Katrina in that no disaster struck. It wasn't catastrophic in Texas for Rita. So, post landfall, it was different. So let us just talk about pre-landfall.

In the State of Texas, they had plans. They had plans at the local level. They had plans at the State level that leveraged the Federal capability. One example, evacuation. In Texas, we were asked—the Federal Government—to provide evacuation support prior to landfall. The State of Texas had a plan to build on, and so we worked with the State of Texas and the areas of Beaumont to Houston to evacuate special needs patients, elderly, and the disabled. One person had an iron lung.

This is a very difficult mission. Just getting one patient onboard an airplane is very difficult, and they did somewhere between 11,000 and 13,000. And we were able to have an effective national response structure—national is local, State, and Federal—because the Federal Government had something to build on. We did not have that in Louisiana.

Chairman COLLINS. Senator Stevens informs me that he has to leave, and he has one question that he would like to ask before he does so.

Senator STEVENS. Thank you very much. I am going to the reconciliation meeting.

One of the problems we are having is that there seems to be a congressional feeling about the extent of our responsibility to the people in New Orleans, who really suffered damage from what I call really man-made disasters. I wonder if the two of you would join me in sending a letter from this Committee asking the Department of Justice if there is a different standard of liability for the Federal Government to those areas that were not harmed by Katrina, but were harmed by the failure of the levee and other systems that were man-made?

Chairman COLLINS. Well, as the distinguished Senator may be aware, we have had one hearing looking at why the levees have failed. We have a second hearing scheduled for next week, which is going to look at all the roles of the various players—the local levee district, the Army Corps of Engineers, and the State of Louisiana, and the Commonwealth of Louisiana's Transportation Department.

There is a lot of confusion, our investigation has found, over who was responsible for the maintenance and the inspection of the levees. I would like to wait until we complete that hearing before proceeding. But I hope you will be able to attend that hearing. I think it is going to be a very interesting one, based on what our extensive investigation has found.

Senator Lieberman.

Senator LIEBERMAN. Yes, thanks, Madam Chairman.

Senator Stevens, obviously, I would be glad to look at any letter you put together. But I agree with the Chairman, it is probably best for us to wait until after the hearing next week.

I will just say very briefly, because I know you have to go, where I think we agree is that the immediate cause of most of the damage in New Orleans was the failure of the levees. And there is some reason to believe—I can't conclude at this point—from what we have heard that there was a failure of construction, design, and other issues.

Where I think we disagree is that, unfortunately, I believe that was foreseeable. In other words, the levees may have failed more quickly and in some different ways than people expected. But we have a lot of history that we put together here that shows that the experts were all saying if a category 3 or higher hurricane hit New Orleans, those levees would not hold.

Maybe more of them broke than we thought. Maybe they broke sooner. But there is a lot of communication indicating they might fail. In fact, the Hurricane Pam exercise that was carried out in June 2004 was based on the levees failing and what would we do as a result. And that is where, I think, we have a reason to ask FEMA why it wasn't ready to deal with it.

Senator STEVENS. Well, I think my question goes to who is going to be responsible? There is a lot of damage out there now that was not covered by insurance, either flood or otherwise.

Senator LIEBERMAN. Right.

Senator STEVENS. And attempts to try and use Federal funds to meet some of that damage is being met with resistance. I think the duty of this Committee is to demonstrate that there is an extra added level of responsibility in the areas where those levees failed.

And it is true that there were predictions. When we did our thing in appropriations in 1998 and 1999, we had the feeling that there were severe problems. We had people that told us there were problems there.

Senator LIEBERMAN. Right.

Senator STEVENS. But notwithstanding that, we got a plan. But no one really fixed the levees, and no one fixed the gates, and people have suffered enormous loss. Now we have to have a greater feeling in Congress about our responsibility to those people who are in that one area where it was not just Katrina, a natural disaster, but damage from man-made disaster.

I think there has been a failure in Congress to recognize that difference. But I appreciate and I hope I can attend the hearing. Thank you very much.

Chairman COLLINS. Thank you.

I am going to resume my questioning with Mr. Wells to try to get us back on track a little bit.

It was helpful to hear your different experience dealing with Louisiana versus Texas. In general, were the Louisiana officials that you dealt with familiar with their responsibilities under the National Response Plan? And did they understand how the Incident Command System worked, in your view?

Mr. WELLS. No. Short answer.

The Incident Command System is very important. You cannot do anything without command and control. If I may take a minute to read something from an emergency services person who is on a workgroup for the search and rescue mission, which was a combined effort of the Louisiana Department of Wildlife and Fisheries, the Coast Guard, the state police for Louisiana, DOTD, and FEMA, to show you what it is like when you don't have an effective Incident Command System? And I am just going to read this verbatim. It is unvarnished.

"Establishing a State/Federal search and rescue workgroup on August 28 was a great concept. However, there were some shortfalls. The workgroup was a joint effort in name only. We had a great initial meeting and established solid operational concepts, but had no decision-makers present with the authority to obligate their agency's resources.

"The State appointed a workgroup leader who was to oversee the State assets. However, this workgroup leader flew to Jackson barracks the afternoon after the initial meeting and was not heard from for 9 days. For 3 weeks, I sponsored"—this was the FEMA person—"the workgroup's twice-a-day meetings that were attended by FEMA, who had resources and authority to direct search efforts.

"The U.S. Coast Guard, who had junior officer representation but no authority to direct search and rescue air operation, all operations were directed by senior Coast Guard officers from another location. These officers refused to meet and conduct joint search and rescue operations with FEMA and State agencies.

"State wildlife and fisheries had representation but no authority to make management decisions on search and rescue operations. That authority remained with senior officers who conducted solo operations.

"State police, who had an interest in following up on 911 calls, but had no State search and rescue authorities or resources to assist.

"Civil air patrol attended and had air resources to support search and rescue, but had no State taskings to engage their resources.

"The DOTD, the State Department of Transportation, had resources to support joint search and rescue missions, but refused to attend any of the meetings or plan joint efforts."

This is a failure of the Incident Command System. It is all about having people that can make decisions on the spot and get on down the road. This is just but one example of why things were slow and why things didn't work out as fast, as efficient, or as effective as they should. If people don't understand ICS, we can't do ICS. And if we can't do ICS, we cannot manage disasters.

Chairman COLLINS. Mr. Parr, what is your assessment of how the Incident Command System worked in Louisiana?

Mr. PARR. I want to echo Mr. Wells's sentiments. It is extremely important, in my experience, at a local level, as something that is near and dear to my heart. I think, and as I said before—and this is not necessarily to point fingers, but hopefully, for us to critique ourselves and learn how we can do better next time—I cannot begin to explain the dedication that the police officers and firefighters in the City of New Orleans, how they acted, how they responded.

The same thing with the National Guard. I was working with people every second of every day who literally lost everything. But they were there doing their jobs, working as hard as they could.

I think one place where we need vast improvement is in their preparation. It is the responsibility of local authorities to evacuate their people. It is the responsibility of local authorities to set up shelters that are properly protected from flood waters, that are properly protected from hurricanes because not all are going to be able to be evacuated. It would be unreasonable for us to expect 100 percent evacuation of a city the size of New Orleans.

I found that there was very little preparation. No information on shelters other than the Superdome. There was no assets or commodities at the Superdome, other than what FEMA gave to the State, which they did distribute at the Superdome.

There were no sanitation facilities before the levees broke. You can imagine the difficulty of moving sanitation facilities into a city with 4, 5, 6, and, in some areas of the city, 14 feet of water once that happened. That is the job of local and State governments. And simply, that just did not happen.

Chairman COLLINS. Mr. Carwile, how did the Incident Command System work in Mississippi?

Mr. CARWILE. Madam Chairman, I think it worked very well at the top. There had been training previous to Hurricane Katrina by the Mississippi Emergency Management Agency on down to the county emergency managers. So it worked well.

We quickly were able to form a unified command with both myself and Robert Latham, my counterpart from the State. And we included at the top of the unified command the Adjutant General for the State of Mississippi, Major General Cross, as well as the commissioner of public safety, George Phillips, because each one of those brought so much to the table in terms of resources to manage.

I think where the difficulty becomes—and I believe Senator Lieberman talked about this in his opening remarks—in a situation, a catastrophic disaster, it is very difficult to build from the bottom up if there is no bottom. Mayor Tommy Longo of Waveland and over in Hancock County, Mississippi, I mean, he lost every fire truck, and the fire station. Every police officer lost their home. There was no city hall. There was nothing left for Mayor Longo to build on.

Similarly, in Hancock County, the emergency operations center in the county had to be evacuated, and we moved our folks over to Stennis in order to have a communications and a coherent system. So I think it is true that it builds from the bottom up. But in a catastrophic, we have got to be able to reach down and to supplement the absence of a coherent system down below.

And to get around that, we predeployed division supervisors with communications from both the Federal and the State teams to the three what we believed would be, most impacted counties with available resources to prop up, if you will, those great first responders and emergency managers and mayors down in the local areas.

We did not do that to the degree we would have liked to have done it because, frankly, there was a paucity of trained personnel to do that. But I think overall, and I believe in ICS. I think that we need to make some modifications as we look at a catastrophic, however, because, to me, the unified command as it is outlined in the National Response Plan calls for unified command only at the top, a few people.

I believe that unified command has to go all the way down the structure, and we have to be able to use State and Federal personnel to prop up local communities that have been totally destroyed by something like Katrina.

Chairman COLLINS. I think that is an excellent point that perhaps we should take a look at whether the system is scalable, whether it makes sense to expect the State and local governments to play the role that is envisioned when the magnitude of the catastrophe may wipe out all of the capability at the State and local level.

Mr. Carwile.

Mr. CARWILE. Yes, ma'am. And I think we can do that without in any way encumbering or impeding the constitutional authorities within the State constitutions of the local elected and appointed officials.

Chairman COLLINS. Mr. Wells, under the current system, FEMA is besieged with requests from State and local governments for various commodities or forms of assistance. I would like to refer your attention to certain exhibits that are in the book that is by you, specifically Exhibits 8, 9, G, H, and I. They all reflect requests made by State and local government entities in Louisiana to FEMA for assistance.

And I am just going to go through what those are. Exhibit 8[1] is from the New Orleans police department asking FEMA for, among other things, 400 M–4 weapons, 25,000 rounds of ammunition, 1,500 pairs of black military boots in various sizes, and 200 Crown Victoria police cruisers.

Exhibit 9[2] is from New Orleans parish, and it asks for 10 gas-powered golf carts to transport firefighters around Zephyr Field.

Exhibit G[3] is from Mayor Nagin's office. It seeks a bus to Shreveport. Exhibit H[4] is also from the mayor's office. It seeks portable air conditioning units to cool offices. And Exhibit I[5] is from the Louisiana Department of Social Services, asking for a taxi to take one person from a hospital to a shelter.

As I was reading through these requests, they struck me as not the typical requests that State and local governments would make to FEMA during a natural disaster. But obviously, I don't have your experience.

First, let me ask you, are these typical of the requests that you would expect to get from State and local governments to FEMA in the aftermath of a disaster?

Mr. WELLS. The problem with these is we got literally hundreds and hundreds of requests like this intermingled with valid requests. And when you get that volume that are not screened, it clogs down the system for legitimate requests that we need to process.

I think this is an indication of just a lack of understanding—this came from the local level—the lack of understanding of what FEMA is there for, what we can do, and probably more importantly, what they can do themselves.

Also, we normally with most disaster operations, when the request comes to the State emergency ops center, they will screen out

[1] Exhibit 8 appears in the Appendix on page 83.
[2] Exhibit 9 appears in the Appendix on page 90.
[3] Exhibit G appears in the Appendix on page 99.
[4] Exhibit H appears in the Appendix on page 102.
[5] Exhibit I appears in the Appendix on page 104.

all of those requests prior to us even getting them. So this is not typical. No, ma'am.

Chairman COLLINS. Thank you. Senator Lieberman.

Senator LIEBERMAN. Thanks, Madam Chairman. Thanks to you three gentlemen.

Having heard your opening statements and the questioning thus far, I go back to what I said at the beginning. You are three extraordinary professionals. I admire the work that you have done. We are lucky to have had you working and to have you still working, at least in two cases, for FEMA.

And what strikes me as particularly significant is that you have something to add from the ground about what we could do to perform better in the next catastrophic natural disaster. That you are proud of your service in FEMA, but you are not defensive about everything that happened. And I think that is the spirit of this Committee. We are not out to get or protect anybody in this. We are out to figure out what happened.

In response to what Senator Stevens said, just beyond what I have said when he was here, there is no question that New Orleans suffered a lot because the levees failed. It was a hurricane that would have had one level of damage. But the levees failing compounded it enormously. The failure of the levees was not only foreseeable, but was foreseen for years if the hurricane was above category 3.

In fact, there is some reason now to believe that maybe the levees failed even earlier with a lower category impact of the hurricane because of some kind of negligence in design or construction. That is a very important question we have to answer.

But having said all of that doesn't mean that FEMA was ready or did everything it should have done. And that is why, in a non-defensive way, we want to get at it. I will say and ask my first questions to you, Mr. Wells, that just a few of the things you have said so far suggest to me changes that ought to be implemented.

For instance, as you said, and in some sense was expanded upon by Mr. Parr and Mr. Carwile in response to Senator Collins' questions, the bottoms-up approach makes a lot of sense in many disasters or emergencies. I wrote it down quickly, so I may have missed it. But when you get to a large disaster, it doesn't work so well. And when you get to a catastrophe, the bottoms-up falls apart.

And part of what I think, therefore, we probably want to do as a government going forward is, as you all said, to put in place for FEMA and the rest of the Federal apparatus some plans in those more significant disaster situations when the bottoms-up won't work, and the Federal Government really has to assume a significant amount of responsibility.

Second very instructive thing I thought you said, you compared the preparation of Texas for Hurricane Rita to the prepareatin of New Orleans and Louisiana for Hurricane Katrina. Texas had an evacuation plan that was adequate to the circumstances. We all understand that Rita didn't hit at the level of catastrophe that Katrina did. But you saw a plan there, which you didn't see in the case of Louisiana.

So this suggests to me that part of what we may want to do as a matter of law or regulation, probably law, is to have some more aggressive Federal oversight of the emergency planning of State and local authorities to the point of having to certify and make sure not just that we look at it, but that they actually have as comprehensive a plan ready for emergencies as possible.

Mr. Wells, let me, in that spirit, go on. Because as I read the statements that you made to our staff in the interviews, I am struck by one quote after another in which you are willing, from the ground, in a very professional and confident way to say this just didn't go as it should have gone. And I am going to quote a little bit.

You talk about FEMA's ability to respond, "But FEMA is not trained. FEMA is not equipped. FEMA is not organized to do very large response operations."

You talk about inadequate communications, "What we had was a communications kind of vacuum here in Baton Rouge."

You speak of FEMA's difficulties in staffing positions. "Just about every position we have, this is a secondary job for people. I think everybody fails to recognize this. Very important."

"Our system is based on—this is whole interagency—who is available at this time, put them in there, get them out the door."

You acknowledge difficulties in planning. "Now we did do some different things here that would need a lot of study like the continuity of government. We didn't work continuity of government at all. In New Orleans, they melted down. Their whole government was just melted down. We didn't have a plan for continuity of government."

And I was struck also by your views of the Stafford Act, "You need different laws. The Stafford Act is not—the Stafford Act is like bringing a donkey to the Kentucky Derby." Have you heard him say that before? "It is not designed for a disaster this big."

So what I am saying is that your candor is very important and very appreciated and very necessary as we work together to try to fix this. And bottom line, it would appear from your statements that FEMA was lacking a plan, communications, appropriate personnel, and various other assets to deal with a catastrophic disaster of this kind.

I want to give you a chance now to comment broadly on those points that you made in the staff interviews.

Mr. WELLS. I don't know if I can remember all that you said, sir. Let us talk about the FEMA part——

Senator LIEBERMAN. Yes.

Mr. WELLS [continuing]. And the Stafford Act and all that. We are not, we do not have the capability versus equipment, people, expertise, training to do large catastrophic disasters. We do not have teams. We do not work as teams. The people you get, and it is not just FEMA—this was a National Response Plan.

And people talk "FEMA this and FEMA that." But you know what? FEMA, once you get out in the field, we are a very small percentage of what is out there. And at the height of Katrina, I think we had, if you include the military, maybe 70,000 people?

Senator LIEBERMAN. Right.

Mr. WELLS. And FEMA was maybe 1,500 or 2,000 of that. And even when the military went down, it was still a small part.

Senator LIEBERMAN. Am I right that full-time FEMA personnel are numbered at about 2,000? This is all around the country, not in that crisis.

Mr. WELLS. Right. There are only 2,500.

Senator LIEBERMAN. And so, what you are drawing on when you say 1,500 to 2,000 personnel for Katrina was redeployment of FEMA full-time personnel from elsewhere, but also a lot of part-time people or people just brought in for the crisis?

Mr. WELLS. Right.

Senator LIEBERMAN. I want to get to that later.

Mr. WELLS. Anyway, we do not have enough people. We rob Peter to pay Paul in disasters. Even in medium disasters, we are doing that. We have 10 regions we have to man. A regional office has to do three things. They have to do two things. They need to do three things, but they have to do two things.

They have to set up a regional response coordination center. It is an emergency ops center. And they have to staff a team that goes forward to Louisiana like we did. Now you only have about 90 people in a region. That is woefully inadequate to do both. You cannot do both. Pick one.

When you get to a field, when we got to Louisiana, we had enough staff for our advance team to do maybe half of what we needed to do for a day shift. We had to do a day and night shift. So we had to prioritize.

We did not have the people. We did not have the expertise. We did not have the operational training folks that we needed to do our mission. And it has been this way for years, sir. Years after years, you are working on the margins. You are getting people from other agencies with no experience or no experience in response operations. They are just filling a billet. We have never trained together.

We need to really train together as a team. We need to work as a team. What you have with this National Response Plan in the field is we have no unity of command.

Senator LIEBERMAN. Right.

Mr. WELLS. FEMA has more of a coordination role. We need to have a command role, where we can direct the sub-elements assigned to us to do things. And it doesn't play out that way, despite what the National Response Plan indicates.

Senator LIEBERMAN. That is a very important point. In some sense, as I read your testimony, as I hear you today, it is as if you are the generals, and you are first rate. But we haven't really given you an adequate trained force to go into battle with you.

I want to ask a few questions similar to Mr. Carwile. I know you also expressed some concern about FEMA's staffing levels and problems. In your opening statement, you talked of how FEMA needs many more trained people.

I know that you had a chance to look at these problems. I wonder if you would describe the impact of funding decisions on FEMA's effectiveness, particularly the ERT–N team. And let me refer you in particular to a memorandum that I believe you contributed to, dated June 30, 2004, which is Exhibit 1.[1]

In it, you refer "to the unpreparedness of national emergency response teams." You say the teams are unprepared because of "zero funding for training exercises or equipment." So I ask you to comment on that in light of what you found in the response to Katrina.

Mr. CARWILE. Yes, Senator Lieberman. This has long been a concern of mine, as well as Mr. Wells.

After September 11, I was asked by Director Allbaugh at the time to reconstitute a national team, an emergency response team national, and to write an operations plan and be prepared to respond to the next terrorist event. It was an all-hazards plan, but the focus, obviously, at the time was on terrorism.

I took a small group from New York City, our joint field office there, went down to Atlanta and put together a team of very seasoned emergency managers and with members of other Federal agencies, including the military, and wrote what was known then as an ERT–N op plan. We were then able to build a team to about 125 individuals, hand picked, from around the country, and we were able to routinely exercise that team because we had the funding in place to do so on the plan against several scenarios.

Senator LIEBERMAN. Were these full-time FEMA employees?

Mr. CARWILE. Yes, sir. For the most part. We did have some reservists, which we called disaster assistance employees, on the team. But the intent was to use them only in certain areas where we have almost no full-time employees to do certain functions within FEMA. We rely solely on our reservists for things like administration and that sort of thing.

We were asked to take the team to the Winter Olympics in Salt Lake City, and we exercised and exercised. And as well as buying satellite communications and being able to, we felt we had a robust plan, a well-trained team, and communications.

It wasn't long after that, and the longer we got away from September 11, the less funding we had. Funding, it appeared to us—and I am just a field guy, not a headquarters person. The small amount of money we did have was being diverted, and we got no money in the out-years. So there was no money.

Then we morphed that team into another team. The red team morphed, and we split it up and formed a blue team. And subsequently, a white team, which was a hollow team. But there was never any money. In former years, prior to September 11, when I was on a national team, we at least had money to do one training event a year and one meeting for the leadership of the teams. That money also went away.

Senator LIEBERMAN. So in that period of years that you have described post-September 11, there was no money for training exercises?

Mr. CARWILE. Except for that first year, sir.

Senator LIEBERMAN. That first year.

Mr. CARWILE. Yes, sir. And after that, the money went away, and the emphasis on the readiness of those teams, as I said in my opening comments, I came from 30 years in the military. We had a very

[1] Exhibit 1 appears in the Appendix on page 72.

rigorous reporting system and the red/amber/green to report unit status. We had nothing like that in FEMA, although those things were being—anyway, I think there was a great problem in resourcing those teams.

Senator LIEBERMAN. Yes. So this was not only a natural disaster waiting to happen, this was a personnel, kind of a FEMA disaster waiting to happen because we weren't giving you the resources to get ready for this.

If I may, Madam Chairman, I just want to ask one follow-up question because I think it is important. It is something I have come to learn myself, in the course of this investigation, which speaks to the fact that FEMA has relatively few full-time personnel.

And this is the response teams and the so-called reservists, which in the interview you did with our staff, it seemed to me you were saying that these teams are mostly names on rosters. They are not really teams because they don't train together, they don't work together, and they don't really have a budget.

And I wonder if you could just give, for the record, a little background on what does it mean to be a reservist for FEMA in this regard? Who are these people, typically? And obviously, I assume you believe, based on what you said before, that if they are there as reservists that we ought to be spending more money training them?

Mr. CARWILE. Yes, sir. I will specifically speak to the fine Americans that form the reserve cadre, which we call the disaster assistance employees, which we rely on almost to a great extent to our ability to surge personnel capacity to respond to any large disasters. In other words, in Florida last year, probably 90 percent of the workforce were disaster assistance employees.

Those employees have traditionally, if you look at the demographics, many of them are retired from all sorts of walks of life. They are people you would be very proud to be associated with. They bring skill sets from decades of experience in various parts of the civil sector and some from the military.

They are inadequately recompensed for the time they spend on active duty. They have absolutely no benefits. None, no benefits whatsoever, even when they are on active duty, we call them up. So, for example, if there is a holiday, and you happen to be on a big disaster, and there is no benefit. If you want to let a few people off, you can't even pay them for being off on the holiday.

Senator LIEBERMAN. Am I right? They don't really train as units?

Mr. CARWILE. Sir, was that the individual part of your question?

Senator LIEBERMAN. Yes.

Mr. CARWILE. The collective part of your question and the teams, there is no training money and very little. Now the regions, as Mr. Wells indicated, they have a responsibility to field an emergency response team for a small to a mid-level disaster. They may, because they are all pretty much co-located, have an opportunity to do some, what we used to call in the Army, collective training as opposed to individual. But there is no money to do team training.

So if you go out on a disaster, to me—and I was very fortunate in Mississippi because I had a team from my home region that I have worked with for years—you need to know how that other person is going to respond in a crisis. You need to have gone through.

In the military, of course, we have a very rigorous exercise program that is evaluated. And I have long advocated, and it was included in a white paper that I wrote last year, that we should have a similar system and have emergency deployment readiness exercises for these teams, and go out and do a rigorous evaluation and give feedback to the team members, much like we do in the military. But that is not done, sir.

Senator LIEBERMAN. Yes. That is a real big gap that we have to figure out how to fill. I assume that there were a lot of people on the reserve rosters who were called up to respond to Hurricane Katrina?

Mr. CARWILE. Sir, one of the national teams was deployed to Louisiana, that is the blue team—the blue team went down. In Mississippi, we had a lot of members that were on the national teams, but we did not deploy a national team, per se. There are only two remaining. Those two teams were reconfigured probably in the last month or so before Katrina.

They were brand-new teams. New members were put in there, some of whom have been on other teams in the past. And they were pared down to, I think, around 25 persons per team, and they, to my knowledge, never had an opportunity to train together beforehand.

Senator LIEBERMAN. Thanks, Mr. Carwile. Thanks, Madam Chairman.

Chairman COLLINS. Thank you. Mr. Carwile, I want to follow up with some additional questions on this June 2004 memo that Senator Lieberman just questioned you about.

First of all, this memo was an extensive memo. It includes many significant recommendations, and it also sounds the alarm. There is the heading on page 3 that says "unpreparedness of national emergency response teams." It says "unprepared teams, zero funding for training exercises and team equipment."

It talks about the need for a single division for response and recovery. It mentions that there had been four different budget proposals submitted over an 18-month period.

And just so we understand, this is from highly trained and important professionals within FEMA. It is the Federal Coordinating Officers. Did you get any response to this memo from Michael Brown or anyone else in his office?

Mr. CARWILE. Madam Chairman, first of all, I wasn't the author of the entire document. But these two——

Chairman COLLINS. You were the author of part of it?

Mr. CARWILE. Yes, ma'am. These two issues were—and I am guilty as charged on that. Ma'am, we put this together, and probably Scott may have had input as well. I am pretty sure he did.

Former Director Brown had asked the Federal Coordinating Officers for their input on things we thought were critical. And my former colleague Mike Hall was elected by the rest of us to put this together, each of us having some input. These were mine.

Mike related to me that this had been submitted to the eighth floor, being kind of the command group in FEMA. And as far as I know, there was never any feedback on any of these issues.

Chairman COLLINS. So, as far as you know, there were no actions taken in response to this detailed set of recommendations that the FCOs sent to Secretary Brown?

Mr. CARWILE. That is correct, Madam Chairman.

Chairman COLLINS. If the recommendations in this June 2004 memo had been implemented, do you believe that the response to Katrina would have been improved? I realize that is speculative. But what is your best judgment?

Mr. CARWILE. Yes, ma'am. I can't help but believe that having trained and effective teams that are well equipped and have had opportunities to work together through training and rigorous exercises would not have made a difference.

Again, I felt very fortunate because many of my colleagues with me in Mississippi had been with me on a national team in years past. It was kind of coincidental.

But I can't help but believe that trained and ready teams, people who have worked together, would not have made some difference in a positive way.

Chairman COLLINS. Could you explain to us why you think that a single division for response and recovery would help improve the response to future disasters?

Mr. CARWILE. Yes, ma'am. And first of all, that is not a novel idea. That was formerly that way. My former colleague is sitting right back here, who was formerly the associate director for FEMA for response and recovery. It was only with the formation of the Department of Homeland Security that the two efforts, that is response and recovery, were bifurcated.

Formerly, there was a greater among equals division within FEMA was response recovery. Because when you go to the field, you don't do response, and then all of a sudden, one day you say, "Well, we are going to quit doing response. Now, guys, we are going to start doing recovery." It is a continuous effort.

I was looking at our timeline in Mississippi. On September 2, we were putting disaster recovery centers out to meet the needs of the people to be able to communicate with us and the Federal and the State officials.

So, to me, and it is also from a person in the field, it is a little bit difficult to know who the heck you work for at headquarters. You know, on some issues, you go to one person. Other issues, you go to another.

When they first bifurcated them, I was in the field, and we would talk to what is now called the National Resource Coordination Center, then known as the Emergency Support Team, and you really didn't know who you were reporting to. Because was it the response guys or was it recovery guys?

So, to me, it is a natural fit. It was an unnatural thing to break it up in the first place.

Chairman COLLINS. Thank you.

I would now like to turn to the issue of evacuation. One of the problems that New Orleans had was evacuating citizens before the storm. Mr. Wells, are you aware of any requests from the City of New Orleans for assistance with evacuation? Any request to FEMA?

Mr. WELLS. No. We would get the request from the State. The city would come to the State, and the State would come to us. And for any pre-landfall evacuation requests, we got none, zero. And as I mentioned earlier, that was totally unlike what we did in Texas, where we were actively involved.

Chairman COLLINS. Now I understand that you were involved in the Hurricane Pam exercise. Did that include a segment on the evacuation of New Orleans and the vicinity? Was that part of the exercise?

Mr. WELLS. No, ma'am. Evacuation was not part of it. When we set up the Pam exercises, we were developing it, we worked with the officials in Louisiana. And they determined I think it was five to eight functions that they wanted us to work as Federal/State partners in it.

And I think we had talked about evacuation, and they said let us leave that off the table because the city and the State, we have been working evacuation issues. And we will park that over there, and we will just work on these other issues. So that was not one of the issues we addressed in the Pam exercise.

Chairman COLLINS. And whose decision, just for clarity, was it to not include evacuation as part of the Hurricane Pam exercise?

Mr. WELLS. Well, it would have been the deputy director of emergency management at the time in Louisiana.

Chairman COLLINS. So it was a State official?

Mr. WELLS. The State. Yes, they determined the issues that we were going to be looking at for Pam.

Chairman COLLINS. Thank you.

Mr. Parr, another issue of evacuation arose after the storm, and that was the evacuation of the Superdome. You have previously told the Committee that you worked with the National Guard at the Superdome throughout Tuesday night and that you had a plan that was actually approved by your supervisors at FEMA and was in coordination with the National Guard to evacuate the victims at the Superdome, beginning Wednesday morning. Is that correct?

Mr. PARR. That is correct.

Chairman COLLINS. Could you explain what that plan was?

Mr. PARR. Sure. As the population of the Superdome started growing almost exponentially on Tuesday and with the waters rising, the breach wasn't able to be closed, we realized we had to get the people out of the Superdome. We felt, the Guard felt that there would be mass confusion, violence once the lights went out in the Superdome. That was the only thing that was keeping people together.

We came up with a plan in conjunction with the Guard. It was the chief of staff or the adjutant general, the ops officer, the commanding general that was there for the Guard. What we were going to do was use Chinook helicopters because they are the largest in the U.S. inventory. The Guard had availability of three.

We came up with a plan to move anywhere from 300 to 500 people an hour out of the Superdome by landing helicopters every 15 to 20 minutes. At the time, we estimated about 15,000 people in the Superdome. We figured we could clear them out within about a day, about 30 hours or less, once we started the evacuation with an additional 9 helicopters.

In addition to that plan, I should say it would be short haul trips. The airport was dry. I believe Belle Chase, which is a base, was also dry. We would have buses meet them there and then take them to shelters after that.

Chairman COLLINS. That sounds to me like a very good plan that would have helped to evacuate people from a situation that was becoming increasingly unsanitary and dangerous. Why wasn't it implemented? In fact, the evacuation did not occur the next day, despite your having what sounds like a very good plan.

Mr. PARR. That is correct. At least the evacuation for the general population didn't begin the next morning, as we had hoped. At some point during the early morning hours—this was a plan we worked on in the overnight hours. None of us slept at all as we developed this and had constant conference calls with Washington, DC.

We couldn't reach our command group at the EOC in Baton Rouge because of communications, but we were able to reach our response and coordination center in Denton, Texas. So we would have conference calls with those two groups.

We were notified at some time around 5 a.m. that General Honoré had taken charge or was in charge of the evacuation of New Orleans and that all plans were to be put on hold, that he would be directing the evacuation. And that was the direction I got from the command group of the National Guard that they would be awaiting his orders.

Chairman COLLINS. So, as far as you know, General Honoré canceled the plan?

Mr. PARR. Well, I can't speak to specifically what happened, but I will tell you that the Guard told me I got a call from General Landreneau at some point probably between 5 and 6 a.m. in the morning, thanking me and thanking us for our hard work. But they were awaiting orders of General Honoré.

Chairman COLLINS. So instead of the evacuation from the Superdome starting on Wednesday morning as it would have under your plan, the plan worked out with the National Guard, when were people actually evacuated from the Superdome?

Mr. PARR. The start for the general population was about Thursday morning, about 24 hours later.

Chairman COLLINS. So the result of the delay of that evacuation plan, which you had worked all night to put together, was that thousands of people in the general population in the Superdome had to spend another very unpleasant, hot, dangerous night in the Superdome. Is that correct?

Mr. PARR. That is correct.

Chairman COLLINS. Mr. Wells, I noted in your bio that you had been the liaison from DOD to FEMA back when you were working for the Department of Defense. DOD has an unusual relationship with FEMA. As I understand it, FEMA can assign other agencies certain missions. But with DOD, it is my understanding the Department has to agree to accept the missions from FEMA, and it is a much longer process. Is that correct? Is my understanding correct?

Mr. WELLS. Yes. When all of the agencies come into the joint field office and we give what we call mission assignments or taskings out to these agencies to do specific things, the approval authority generally rests with the person in the joint field office, and it gets done immediately.

But the Department of Defense, their approval authority rests with the Secretary of Defense. And so, it has to go through a long process of validation and through their chain of command to get it approved. And that is more than awkward. It is more than cumbersome. It just takes a long time to execute.

I need to say, parenthetically, that in Katrina we did not see that lag that we normally see in most disasters, and they were fairly responsive.

Chairman COLLINS. But from your perspective, since you have seen it both as a DOD employee as well as a FEMA employee, should the Department be treated differently?

Mr. WELLS. No. Having DOD is sort of like somebody giving you an 800-pound gorilla. You are supposed to take care of that gorilla and be responsible for that gorilla. But that 800-pound gorilla is going to do what he wants to do, when he wants to do it, and how he wants to do it. So you lose some of that control in your organization with the Department of Defense structure.

What they have is, when General Honoré came in, for example, he had really two organizations. He had a defense coordinating officer, who was in that joint field office. And he had a brigadier general, Mark Graham, with the staff, who worked directly with us out of the joint field office. And that fell within the architecture of the National Response Plan, and it was more organized.

General Honoré had a joint task force that went and did things separate and beyond that. He did great things. Him and his joint task force did great things, but it wasn't coordinated, and it led to some problems.

Chairman COLLINS. Mr. Carwile, did you have something you would like to add to that?

Mr. CARWILE. Madam Chairman, in a previous life, I served as a defense coordinating officer, as Mr. Wells talked about, and I have had discussions with Secretary McHale on this very subject. And we spoke earlier about a unity of command and a unified command, and what Mr. Wells just described is outside of unity of command, unity of effort, and a unified command.

In other words, you can't have two Federal agencies, even if one is an 800-pound gorilla, operating independently of other Federal agencies. And there is a difference of opinion of my personal opinion as a private citizen between what I read in the Stafford Act and the way that the current secretary and assistant secretary for homeland defense read their authorities.

And as Mr. Wells indicated, we mission assign other Federal agencies. DOD, I personally know—Secretary McHale and I have discussed this—takes exception with even the term "mission assignment" from FEMA to the Department of Defense.

Chairman COLLINS. Mr. Parr, do you have anything to add to that debate?

Mr. PARR. I can tell you on a local level. After leaving New Orleans, I worked in St. Bernard Parish for about 30 days. I think Mr. Wells' description of the 800-pound gorilla, because we had a major significant military DOD presence there, was true.

They did some great things. The men and women that were there were truly phenomenal. But keeping them—and I will use this term, but it is not pejorative—reined in to keep them in—remember, when they were there on the ground, it wasn't me that was directing operations in the parish. It was the parish leadership. It was the parish president. It was the sheriff. It was the parish emergency manager.

And keeping them from running over that was my job because I was supposed to be the lead Federal person there. And it is difficult. But they do bring a lot to the table, and they are a necessary part of the National Response Plan.

Chairman COLLINS. Thank you. Senator Lieberman.

Senator LIEBERMAN. Thanks again, Madam Chairman.

The more we go on, the more I keep coming back to military terms because, in some sense, you are in a real crisis, almost a battlefield situation. And one I want to talk about is situational awareness.

Obviously, in military matters, we are, through technology, dramatically improving the situational awareness of our forces. And that is not a bad ideal to set up for emergency response here as well. But it was, really in this case, lacking, again beginning with the fact that this was an extraordinary disaster. But part of what I want to get at is the details of that.

Mr. Parr, I am going to ask you the first questions. We have seen photographs and video showing the absolutely devastating situation at the convention center. I know that most of the accounts have focused on what went on at the Superdome. But I want to ask you to talk a little bit about what was unfolding a mile down the road at the convention center, where thousands of recently homeless people sought refuge from the storm.

I have taken a look at your notes from August 30, which is Exhibit F, and they mention that you had a briefing with Bill Lokey and Scott Wells. And your notes mention the convention center. But we don't see anything in your e-mails or any other documents from yourself or the FEMA team regarding the circumstances in the convention center on Tuesday, Wednesday, or Thursday.

So I wanted to ask you during the briefing with Mr. Wells and Mr. Lokey, first, what did you discuss, to the best of your recollection, regarding the Superdome?

Mr. PARR. OK. You have an advantage over me, Senator, because since this was kind of a last-minute request for me to appear before you, I left my notes in Texas, where I am assigned right now. So I am not able to look and see exactly what you are referring to.

Senator LIEBERMAN. I am glad to try to get our staff to give it to you.

Mr. PARR. I think it is Exhibit F?

Senator LIEBERMAN. Yes, it is Exhibit F[1] in that exhibit book.

Mr. PARR. Just one second. I am sorry.

Senator LIEBERMAN. You went to the same handwriting course that I did.

Mr. PARR. You noticed that, sir.

Senator LIEBERMAN. It is on page 8, noted at the bottom, 0008.

---

[1] Exhibit F appears in the Appendix on page 94.

Mr. PARR. Let me address, I believe, the phone call you are speaking about because this was probably the only time we were able to get a hold of the EOC, where Scott Wells, the deputy, and Bill Lokey were.

Senator LIEBERMAN. So they were in Baton Rouge?

Mr. PARR. Baton Rouge. Correct. And this was probably the only time we had any conversation, certainly any extended conversation with them.

Senator LIEBERMAN. And at that point, you were——

Mr. PARR. I was? This was Tuesday. I would guess early afternoon?

Senator LIEBERMAN. Right.

Mr. PARR. Or mid afternoon?

Senator LIEBERMAN. And you had arrived a little earlier that day.

Mr. PARR. We had arrived at some point Tuesday morning or late morning.

Senator LIEBERMAN. Right.

Mr. PARR. This is when we went from the Superdome to the city EOC, which was across the street. At that point, we had called back to the State EOC. The primary purpose of that call was to give the city's list of priorities to the EOC there up at the State, for FEMA and the State to start working on those priorities.

If the call was an hour, the biggest part of that call, probably 45 or 50 minutes, was getting visibility on what was being done to close I believe it was the 17th Street Canal breach.

Senator LIEBERMAN. Right.

Mr. PARR. That is what was filling up the city, and there was a whole domino effect of things that happen if that breach was not closed as quickly as possible. Just to mention two of those things. The power plant, once that went under water, and it is my understanding that it was within inches—2 or 3 inches of going under water—the city would take about 6 months, it would probably still be under water right now——

Senator LIEBERMAN. Yes.

Mr. PARR [continuing]. If that was submerged. The emergency power at the Superdome was literally within 1 inch of overflowing the sandbags that were protecting the emergency generator there. We would have lost power there. It would have caused chaos there.

So gaining visibility on that. Talking about it was the city—I am not sure of his exact title. I believe he was either the emergency manager or the homeland security person.

Senator LIEBERMAN. Let me interrupt you. By "gaining visibility," you mean trying to get attention from the EOC onto those two significant problems you have just talked about?

Mr. PARR. I was informed by Colonel Terry Ebert, who is the city's, I think, homeland security director, that the breach——

Senator LIEBERMAN. Right.

Mr. PARR [continuing]. That the Army Corps had stopped attempting to close the breach, and we weren't sure why. And there were some significant issues with them trying to close the breach, and they were unable to do it, and they were working the problem. We did not know that at the time, which is one thing that prompted the call.

So getting that back up on the table and letting them know back up at the State why it was so important to get that closed. Which that did not happen, at least not at that point, was the primary purpose of that call, and in addition, the other priorities that the city had, the Colonel, Terry Ebert, gave it to them on that call.

Senator LIEBERMAN. Yes. Did you become aware in that conversation of the problem of the growing crowds at the convention center?

Mr. PARR. Let me say that we were helicoptered into the Superdome and had no movement, no visibility of the city. The only thing we had was what the National Guard had given us. I don't quite see my note on the convention center. I am sure I was told by the Guard that there were issues at the convention center.

Like I said, I don't have my notes. But because we were, in effect, literally on an island or stranded at the Superdome, if I have a note here about the convention center—OK, yes. I don't think, that was not a note about conditions at the convention center. I see it now. I am not even sure I was aware of conditions at the convention center at this time.

What that was, if you notice it says "convention center, EOV or com suite," that was mentioned as another place that we can carry out operations because of the difficulty of carrying out operations at the Superdome because of the situation there. I was trying to identify another place in the city where we can operate from, and that was a place that was mentioned.

Senator LIEBERMAN. Yes. So, obviously, you were at the Superdome. So you had an awareness of what the conditions were there?

Mr. PARR. Yes, sir.

Senator LIEBERMAN. But you did not know at that point what the situation was at the convention center?

Mr. PARR. Correct. There would be snippets from certain Guardsmen who were patrolling the city saying that the convention center was filling up. I am not sure when or how that was brought out to me. We had no personal knowledge of what was going on at the convention center.

As a matter of fact, I learned more about what was going on at the convention center when I left and went back up to Baton Rouge. Then I had visibility on it.

Senator LIEBERMAN. Yes. So during that period of days before you went back to Baton Rouge, which, if I recall correctly, was Tuesday, Wednesday, and Thursday of that week?

Mr. PARR. Correct.

Senator LIEBERMAN. The FEMA teams that you were involved with did not do anything with regard to the convention center?

Mr. PARR. No, sir, and I think this is an important point that we need to remember, and I alluded to this earlier in one of the answers to one of the questions that were given. There were no, to my knowledge, identified shelters in the City of New Orleans.

Senator LIEBERMAN. Right.

Mr. PARR. That is a very important thing. What was used as a shelter was any place that was dry. Any overpass, any high piece of ground. It is important that the city had identified shelters and areas of refuge, and they did not.

So I think that was an ad hoc—since it was largely dry, it is my understanding—an ad hoc shelter that people flowed to simply because it was out of the pool of water in the city.

Senator LIEBERMAN. Thanks, Mr. Parr.

Mr. Wells, you were a senior Federal official at the emergency operating center in Baton Rouge. When did you become aware of the crowds and the problems at the convention center in New Orleans?

Mr. WELLS. To be candid, I am not exactly sure. It was probably around I would say Wednesday or Thursday, we got calls from here in Washington, DC. "What is going on at the convention center?" And I think they were getting reports from the media.

Senator LIEBERMAN. Yes.

Mr. WELLS. And as I recall, we tried to get communications up with Phil and his folks. We couldn't get communications with him to have them go check it out. And so, we went to the State and asked them what they knew, and they used their National Guard and State police assets to give us some visibility. And the big issue was how many people were there? And what is going on, and how many people were there?

And we did not get a clear picture. It probably took 24 to 48 hours from the time we started asking the question to get a picture of just what was going on in the convention center.

Senator LIEBERMAN. So your answers, I think, illustrate the real problems there were in communications under the circumstances there and the inability to have anything approaching the kind of situational awareness that you would want to have or we would want you to have.

I want to go back to something you said in your interview with our staff, Mr. Wells, that there was "a big communications void," which created a black hole in communication abilities from the emergency operating center in Baton Rouge and New Orleans. And if I asked you why, I know it may be an obvious question. But for the record, since obviously FEMA is supposed to be prepared for emergencies of this kind, why did that black hole of communications occur?

Mr. WELLS. I don't know. I was told all of the lines were saturated. The big vacuum was in Baton Rouge. No one could get in to us, and we couldn't get out to anybody. The people in New Orleans could talk to our regional office in Texas in a degraded way, and they could talk to the national office here in Washington, DC. But nobody could call us in Baton Rouge, and we couldn't call out. So that was the biggest vacuum in Baton Rouge.

Senator LIEBERMAN. Mr. Parr, I know that at some point in your testimony or work with the staff, you described FEMA's Red October, the mobile command post 12 feet longer than a tractor-trailer with the most elaborate state-of-the-art capabilities for communications, which would have served as the base for communications in New Orleans.

As the vehicle was in transport to the Superdome, I would just say for the record, it was determined that you could not get it into the area, and you were left with no communications at the Superdome. Am I right about that?

Mr. PARR. That is largely correct, sir. That was the original intention to either move that vehicle or a similar vehicle into the Superdome, but because of the water around the stadium, we could not get it in.

Senator LIEBERMAN. And there was no backup plan for anything else in that kind of circumstance?

Mr. PARR. Well, sir, initially the backup plan was to have a smaller vehicle, a vehicle that was a little bit more maneuverable get into the stadium. I don't believe that the situational awareness allowed us to know. I don't believe that it was common knowledge. It certainly was not my knowledge that water would prevent us from getting those vehicles into the Superdome.

Senator LIEBERMAN. Right. And this is the final question because my time is up. You did previously state, as I mentioned and Senator Collins did, that your access to communications was "extremely, extremely limited" and affected your operations effectiveness, what you were able to do in the circumstance by, in your opinion, about 90 percent. Is that right?

Mr. PARR. Yes, sir. And just to expand on that a little bit, as I look back on the events that happened, we were able to achieve a lot. At the time, it was extremely frustrating. It might take 2 or 3 hours to get through to the people we were calling, but we did get through and we did get things done as far as the evacuation of the Superdome. We made sure that we kept some eye on visibility of commodities so people ate.

But I believe we could have accomplished a lot more if we had the proper communications.

Senator LIEBERMAN. Sure. And when you did get through on the calls, what were you using?

Mr. PARR. The National Guard had two communications vehicles, one with one phone and one with I think three or four phones. So it was kind of like waiting in line. When they finished their business, then we got to go.

Senator LIEBERMAN. Thanks. Thanks, Madam Chairman.

Chairman COLLINS. Thank you.

Just a couple of concluding questions. First, I realize all of you were deployed to the Gulf region in your roles as FCOs. But are any of you living in the Gulf region? Mr. Wells.

Mr. WELLS. I am staying there.

Chairman COLLINS. Pardon me?

Mr. WELLS. Yes, I will be in Baton Rouge.

Chairman COLLINS. You are now. But prior to being deployed there, where were you living?

Mr. WELLS. I was living in Texas.

Chairman COLLINS. Thank you. Mr. Parr.

Mr. PARR. I have only been living in the Gulf since August 28. But my home now or at least where I rent an apartment is in Boston, Massachusetts.

Chairman COLLINS. I noticed you were Region 1, Senator Lieberman's and my region.

Mr. Carwile.

Mr. CARWILE. Ma'am, my family is from Gulf Shores, Alabama. It is in Baldwin County, Alabama, on the coast. They have evacuated probably six times in the last 2 years, with my mother, who turned 89 last month. I personally reside in the State of Hawaii, though.

Chairman COLLINS. I was surprised, when I looked at where FEMA officials were deployed from, that they weren't individuals already assigned to the Gulf region. Would it increase the effectiveness of FEMA officials if they were from the region to which they are deployed, or does it not make any difference?

The reason I ask is there have been some indications that FEMA officials were just not very familiar with New Orleans or the areas to which they were assigned. And I am wondering if it would be better to have the FCOs come from the area or whether it doesn't matter. Mr. Carwile.

Mr. CARWILE. Madam Chairman, the Congress authorized FEMA 25 Federal Coordinating Officer positions back in 1999.

Chairman COLLINS. Right.

Mr. CARWILE. And those positions were sprinkled throughout the country. And ideally, the Federal Coordinating Officer would come from the home region impacted. For example, if the State of Maine were impacted, hopefully, Mr. Parr would be the person, as he would be.

What happens in a very large disaster like Katrina, the resources are quickly stripped down in terms of personnel who are familiar to the area. So over in Alabama, Mike Bolch, who was the Federal Coordinating Officer, did come from Atlanta in the home region. But FEMA can get very quickly overwhelmed in terms of having people with regional experience.

And Mississippi, fortunately, the team from Region 9, which was California and the West, that we brought in had been with Mississippi during Hurricane Dennis and also with a short exercise we had prior to the hurricane season in Washington.

Chairman COLLINS. Mr. Parr, do you have anything to add to that?

Mr. PARR. I will say all of us are national assets, and traveling is very demanding on all of us. I know in the 2 years I have been in the Federal Coordinating Officer program, I have traveled about 300 days.

Chairman COLLINS. Wow.

Mr. PARR. I don't know. I think it is important that we have significant representation from the region that is experiencing the disaster. So, for instance, Scott Wells is a resident and member of Region 6, which is a large part of the Gulf Coast.

The fact that I come from Region 1 and have a lot of experience in Region 2, which would be the New York area, I don't know is significant. I think all of us are kind of, to a large extent, plug and play. You bring us to where we need to go, and the actions that we have to do are pretty much the same.

Chairman COLLINS. Mr. Wells, are you from the region?

Mr. WELLS. Yes. And I think there is value added. You get your value in the peacetime planning. For example, with Katrina, prior to Katrina, in July, just a month before, Hurricane Dennis was threatening Louisiana, and we deployed our response team to Louisiana and did some training with them. So we had built up some experience with the State and did some hasty planning and things like that. And that was a benefit when we went back for Katrina.

Chairman COLLINS. Thank you.

Mr. Carwile, my final question is for you. Both Senator Lieberman and I have been to the Gulf region since the hurricane struck. The damage in New Orleans and Louisiana is incomprehensible. But what really astonished me was the devastation in Mississippi. It is just extraordinary.

Do you think Mississippi has gotten the attention that it needs compared to Louisiana? I am not trying to play off one State against another. That is not my purpose. But the devastation from wind damage in Mississippi matches in many ways the horrendous devastation from water that we saw in New Orleans.

Mr. CARWILE. Madam Chairman, I have been on, I think, every major disaster in the last 9 years with FEMA to include four hurricanes last year in Florida and a number of super typhoons in the Pacific, and I have never seen—and I had two combat tours in Vietnam, special forces. I have never seen the damage that I saw and you saw in Hancock County and Harrison County. Total devastation of entire communities.

We talk about communications. I have never been in a situation where we had such a shortfall in communications. Last year in Florida, fundamentally, we could use cell phones. Practically, just most of the time, we had to rely on sat phones. But I am talking about no communications.

Senator Lieberman talked about situational awareness. Very difficult to have down there other than overflights. I know that Governor Barbour and his staff in the State of Mississippi have done an extraordinary job of leadership. I do think that there has been an awful lot of focus on the visual on New Orleans, and obviously, there was a great deal of suffering there that needed to be tended to.

I do think that there could be more attention paid to the restoration of the Gulf Coast of Mississippi, especially. I say "Gulf Coast," but I was in Jackson when a storm—we had a category 1 hurricane go over Jackson. We were without power in Jackson for days and days, all the way up to the northern part of Mississippi.

So the visuals are on the Gulf Coast, and obviously, that is terrible. In addition to wind damage, the 30-foot surge just cleared off whole counties practically. It was total devastation. But I would concur if someone were to suggest that Mississippi probably deserves more attention than it has been getting, ma'am.

Chairman COLLINS. Thank you. Senator Lieberman, do you have any questions?

Senator LIEBERMAN. I do, a couple. I will try to do them quickly. Thanks, Madam Chairman.

I want to come back to the discussion about the communications and situational awareness, and I am struck that the National Guard had those two communications vehicles in the Superdome. And I understand this is always a question of hindsight is clearer than foresight.

The understanding was growing in the preceding week that this was going to be a hurricane category 4 or a category 5. I was going to read this to Senator Stevens, but I didn't.

There was a communication from William Lundy of FEMA to you, Mr. Carwile, and to Robert Fenton and others, e-mailed on Saturday, August 27, at 11:41 p.m.

"Recent recon flight reports category 4 now, maybe a weak category 5 by Monday morning. Landfall around noon. Storm is carrying a lot of moisture. Experts predict that the levees protecting New Orleans will be breached late Sunday night, thereby flooding the city to a depth of 6 to 12 feet. Storm surge 13 feet, with 20-foot waves on top of that."

Unfortunately, that e-mail was pretty much correct, except it estimated a little bit earlier than the levees actually broke. So here is my question. Again, hindsight clearer. And this is why some people say FEMA moved slowly. You have got this Red October, a fantastically equipped communications vehicle—why not move it into New Orleans in advance of landfall so you are there with a communications apparatus before, what was being talked about and predicted, the levees broke and the city was flooded?

Yes, Mr. Parr?

Mr. PARR. This was something that I discussed with the staffers that were down beforehand. One thing that you never want to do, especially when you are a response—we are not a rescue organization—but a response organization or an emergency organization, is put equipment or people into harm's way.

If we had brought those vehicles into the Superdome, they would have been exposed. They probably, almost certainly with the high winds of a hurricane hitting, would have been damaged and rendered useless. There are other things that we could have done. And in hindsight, if we had been fully aware of the situation, we could have helicoptered communications suites in.

Senator LIEBERMAN. Beforehand, you mean?

Mr. PARR. No. Afterwards.

Senator LIEBERMAN. Afterwards.

Mr. PARR. But in my opinion, and there is no response organization that I am aware of that puts people or equipment in harm's way prior to landfall of a hurricane, unless it is a rated hurricane shelter beforehand, sir.

Senator LIEBERMAN. Yes, I have learned this as we have gone on, and I am not prepared to argue against it. But I am prepared, not now, to question it as we go along.

Because, obviously, the people on the ground there, the local fire, police, emergency personnel are there. And if you see something this big coming, it is tough. I don't have an easy answer to this one. But it seems to me that there is an argument to be made that you would want to try to the best of your ability to get some people in, maybe equipment in there beforehand. But I want to come back to that.

Mr. Parr, I do want to give you, while you are here, an opportunity to answer questions that were raised in Mr. Bahamonde's testimony and then in other testimony before the Committee about the fact that you and your teams left the Superdome on Thursday. You were there Tuesday, Wednesday, Thursday, and then left on Thursday, which left, obviously, the Superdome without any FEMA presence, including the medical team that you had.

So, obviously, we ask why. And I want to give you the chance while you are here to answer that.

Mr. PARR. Certainly. I could give you the long answer or the short answer. So I will try to give you something in the middle ground that would give you the situation.

Senator LIEBERMAN. OK. You have got two centrists up here. So—— [Laughter.]

Chairman COLLINS. The middle position is always a good one.

Mr. PARR. I think that we have described the situation in the city as a whole. The situation in the Superdome was always a powder keg. The Guard first made the suggestion, one of the generals of the Guard first made the suggestion that I consider leaving Tuesday night.

Senator LIEBERMAN. Do you remember which one that was?

Mr. PARR. Yes, sir. It was General Veillon.

Senator LIEBERMAN. General Veillon.

Mr. PARR. And that was when the city was filling, and they felt that once the lights went out that there would be pandemonium there.

Senator LIEBERMAN. So was he suggesting that your safety might be in danger?

Mr. PARR. Well, I believe the school of thought was everyone, including the Guard—because, remember, not everyone in the Guard was armed. Most of the Guard was involved with getting food, commodities in, running and helping with medical missions. As I said before, they did a phenomenal job in search and rescue, in everything, and assisting the New Orleans PD as much as they possibly could. So I think it was everyone's safety.

To skip many things, Thursday morning at first light, General Jones, who was the commanding general in place, said to me—I don't remember his exact words, but there are certain phrases that he said that stick out in my mind very clearly. "I don't believe I can protect you or your people any longer. We are going to be making our last stand," and he pointed to a portion of the parking lot over there. He says, "Get behind us, and we will do what we can."

That is when we started making plans to leave. I spoke, since it was my responsibility to take care of the Federal forces on hand, I spoke to the NDMS teams, told them to continue to operate. I wanted to have helicopters standing by if the situation degenerated. We were unable to get helicopters in immediately.

To make a long story short, I was informed by the DMAT team, the medical assistance team that was in the basketball stadium, that the Guard, in shortening their lines, had pulled all security from them. They told me that they did not feel safe and that they were evacuating. And they had high-water vehicles that they used to resupply and that they were pulling out in their vehicles.

It was only at that time that I made the decision to leave since if conditions did degenerate, we would have no other way out.

Senator LIEBERMAN. I want you to know that some of the other folks who were there say that, in fact, there was not any behavior that would lead them to think that there was a riot or that there was a safety problem. I guess my final question to you is the following—to what extent did you see any behavior that would lead you to think that your personnel would be in danger, or was it derived from what the National Guard folks told you?

Mr. PARR. There are many things that I left out in not giving the answer in total. The intelligence that we had all came from the Guard. That is the first thing.

The second thing is is that—I just lost my train of thought. But one of the things to remember, when I did make the decision to leave, is not only did the Guard say that. You could see from some of the memos that the city asked for with 400 rifles, etc., the Guard—and this was what I wanted to say. The Guard had intelligence that there would be riot. That people would move in force against the Guard at some point that late morning.

Senator LIEBERMAN. And they told you that?

Mr. PARR. Yes, sir. And one of the things to remember is that I have worked in riot situations in my time as a firefighter and fire officer in New York City. I have worked in near-riot situations. The time for unarmed people, that is specifically a security and law enforcement issue, and the time for unarmed people to leave is before a situation starts, not after. And that is when I made the decision to leave when I did.

Senator LIEBERMAN. OK. I appreciate your answer.

We are running out of time. I am going to mention something else. Maybe I will send a letter to each of you and ask for your response.

This goes also to pre-storm, and the question is the following—in a circumstance like the one we saw coming here, a category 4 or 5 hurricane, with potential talked about widely that the levees were going to break. And what seemed to you, Mr. Wells, to be an inadequate evacuation plan in Louisiana as compared to what you saw in Texas, should we in the future have the Federal Government, perhaps through the military, be prepared to do pre-storm evacuation?

Let me state it in a dramatic metaphor. If we had intelligence that led us to believe that a bomb was going to explode in one of our major cities within 3 days, and if we thought it was accurate intelligence, but we hadn't found a bomb, I presume we would use whatever Federal resources we had to get in and evacuate as many people as we could.

And in some ways, though not quite the same, if you put all the facts together, we were in a somewhat similar situation with Katrina. And my question is should the Federal Government try to develop a kind of standby capacity, particularly using the military, and have the ability to assist in that kind of massive evacuation?

Anybody really want to give a quick answer? Sorry.

Mr. WELLS. The answer is yes. And Katrina clearly showed that. But it needs to be expanded. I mean, the continuity of government. We need to have that capability. But this bottoms-up approach only works to a certain point, and we need to have a Federal capability that once you get beyond that point that can make up the difference. And it may mean a totally different architecture.

Senator LIEBERMAN. Right.

Mr. WELLS. If that is the way it is, so be it. But this evacuation is a very important thing. And if we would have gotten those people out in time, we wouldn't have lost close to 1,100 lives in Louisiana.

Senator LIEBERMAN. Absolutely right.

Mr. WELLS. So those are things that I think we need to look at to make it better next time so we don't have this happen again.

Senator LIEBERMAN. Mr. Carwile.

Mr. CARWILE. Senator Lieberman, if I might? I agree with my colleague, Mr. Wells, on this point in terms of capacity. I get in a little trouble, I think that the governors of the States have constitutional authorities that we can bring Federal capacity, whether it is military or other, to bear in support of them to whatever degree they are comfortable with. But I just throw that small caveat out.

Senator LIEBERMAN. Thank you. Mr. Parr, do you want to answer?

Mr. PARR. Yes, sir. I also agree. I would defer a little bit. I think the military, the DOD is a support group in every emergency support function we have because they have so many talents and bring so many expertise to bear. In my opinion, I do not think they should be the lead in an evacuation, but certainly their assets could be used to help an evacuation.

It might actually mean changes in statute. Until there is a disaster, the Federal Government has limited involvement, at least until there is some sort of declaration. I would like to see, personally, and I believe Mr. Wells in his writings has talked about expanding FEMA's role pre-disaster declaration. So that we can, even if it is just technical assistance, provide some assistance to States and locals from the Federal perspective in helping for evacuations.

Senator LIEBERMAN. Very helpful responses. Thank you. You have been very helpful witnesses overall in assisting the Committee in fulfilling its responsibility.

Thanks, Madam Chairman.

Chairman COLLINS. Thank you.

I, too, want to thank each of you for appearing today, for your candid and insightful testimony. It has been very helpful for us to hear from those who were involved directly in the operations.

We are going to be continuing our investigation and ultimately drafting some legislation for reforming the system, as well as recommending administrative reforms. And I would invite you to keep in touch with the Committee, and I hope you will be willing to react to proposals as we go along because you do have so much experience that I think is very helpful to this Committee as we attempt to determine what went wrong and what reforms are needed. So I thank you very much for your testimony and your cooperation.

The hearing record will remain open for 15 days. We will include your complete written statements in the record, as well as any other materials that you wish to submit.

This hearing is now adjourned. Thank you.

[Whereupon, at 12:20 p.m., the Committee was adjourned.]

APPENDIX

SCOTT WELLS

Federal Coordinating Officer,
Louisiana Hurricanes Katrina & Rita

TESTIMONY

BEFORE THE

Senate Homeland Security and Governmental Affairs Committee

ON

HURRICANE KATRINA: PERSPECTIVES OF FEMA'S OPERATIONS PROFESSIONALS

DECEMBER 8, 2005

Good morning Chairman Collins and members of the Committee. My name is Scott Wells, and I am honored to appear before you today. My current position with FEMA is Federal Coordinating Officer (FCO) in Louisiana for Hurricanes Katrina and Rita. I would like to start this morning by thanking you for the invitation to testify before this committee. I appreciate the opportunity to share my perspective on the FEMA operations before, during and after our nation's costliest disaster, Hurricane Katrina. It is my intention today to speak candidly with you about my experiences in Louisiana, both leading up to and following Hurricane Katrina, as well as about my perspectives on emergency management. The views expressed in my testimony are my own and do not necessarily reflect the views of FEMA or the Department of Homeland Security.

I will begin my testimony today with a brief overview of my professional career in emergency management.

For almost two decades I have served in various emergency management positions. Beginning in 1985, I served for two (2) years as a first-responder MEDEVAC pilot. During my twenty-four year military career, I also spent ten (10) years in the Pentagon providing military support to civilian authorities. My last military assignment in the Pentagon was as the Military Liaison Officer to FEMA. In these assignments I was involved in numerous disasters and emergencies such as Hurricane Andrew, the Northridge Earthquake, the Midwest floods of 1993, the Oklahoma City Bombing, the Haitian/Cuban Immigration Emergency, the Waco Siege, and the Ruby Ridge incident.

I retired from the Army in 1999 and have been working for the Federal Emergency Management Agency as a Federal Coordinating Officer (FCO) for the past 6 years. I have served as Federal Coordinating Officer for several disasters and emergencies throughout the United States including Tropical Storm Allison in Texas and the Columbia Shuttle Recovery Operation.

On 27 August, 2005, I was assigned to Louisiana as the Deputy Federal Coordinating Officer to Bill Lokey for Hurricane Katrina. I served in that capacity until 19 September. At that time, I was reassigned to Texas as the Federal Coordinating Officer for Hurricane Rita, as she made her way through the Gulf of Mexico. As the FCO for Rita, I remained in Texas until the first week of October, at which time I returned to Louisiana to replace Bill Lokey for Hurricane Katrina.

Detailed preparation for a Katrina landfall in Louisiana started in earnest on Friday, 26 August when the National Weather Service quickly changed the projected zone for landfall to include Louisiana. Much work had been done earlier in the week, but the focus of those efforts—given the projected path of the storm--was on Florida, Alabama, and Mississippi. Team deployments to Louisiana began on Saturday, 27 August, with the Emergency Response Team- Advanced (ERT-A) and Emergency Response Team-National (ERT-N) elements deploying to the Louisiana Emergency Operations Center in Baton Rouge. The federal regional and national staffs consolidated that night and started conducting field operations in preparation for landfall. In addition to the command and control element being set up in Baton Rouge we concurrently were setting up an Operational Staging Area in Alexandria, Louisiana, that served as a federal logistics base for Katrina operations.

The first actual employment of federal resources—that is, where we provided response assets to the state--occurred on Sunday when we shipped 6 truckloads of water and 3 truckloads of MREs to the Superdome. Two of the truckloads (one each water and MRE) did not complete delivery. They were denied entry by the Louisiana State Police before reaching the Superdome. This was soon followed by many other response resources such as medical teams, search and rescue teams, and additional critical commodities such as water, food, and ice.

There has been much said about the slow federal response to Katrina. From my perspective I want to say nothing could be further from the truth. We had a fully operational logistics base, a fully operational command cell, and response teams in place—all before landfall. We even moved some supplies in before landfall and attempted to move in a medical team. On the day of landfall we moved search and rescue teams, medical teams, and critical supplies into the affected area. That was fast; it may not have been enough for an event of this magnitude, but it was fast.

I think the real issue is that the response was not robust; it was not enough for the catastrophe at hand. As you look—as we all look to make it better next time—I think it's an important distinction to make. Slow means one thing; not enough means something else. The corrective actions between fixing "slow" and "not enough" could be significant.

Emergency management is unlike any other system in the government; it is a bottoms-up approach; the people on the ground are in charge. The first responders are supported, as required, by local government, then state government, and, as a last resort, the federal government. Ultimately, authority for disaster response operations rests at the local level. The state and federal governments are not in charge but are responsible for assisting local governments. And that is how it should be as all disasters are local—disasters start at the local level and disasters end at the local level. This system works for small to medium disasters. It does not work so well for large disasters, and it falls apart for a catastrophic disaster. I think that is a fundamental problem with the response to Katrina.

Following are some of the other major problems and proposed changes I believe could improve our national—local, state, and federal –readiness posture to respond to future disasters.

Need to strengthen the emergency management capability at the state and local level.

- There is very little capability at the state and local level. States generally have not allocated sufficient resources for a viable capability for medium-to-large disasters much less a catastrophic one.
- State/local communities generally do not have on hand--ahead of a disaster--adequate critical supplies and equipment such as water, food, generators and communications equipment to be self sufficient for the first 72 hours following a disaster. This results in the federal government taking on a "first-responder" role in emergency management, which is outside the scope of the federal mission.
- Local/state emergency management staff size (and expertise in many cases) is generally inadequate to perform the critical functions during response operations. This also applies to pre-disaster planning and preparedness activities. States do not have sufficient staff and resources to accomplish the requisite planning/preparedness activities to attain a viable readiness posture for a large to catastrophic disaster.

Need to review the emergency management architecture for response and recovery operations. There are problems associated with the implementation of the Stafford Act as it is executed through the National Response Plan (NRP) and the Incident Command System (ICS).

- The emergency management community at the state/local level has not embraced ICS. The level of acceptance and understanding of ICS is highly situational depending on the state. States like Mississippi embrace the concept, and it worked there for Katrina. Unfortunately, in Louisiana, there was little to no understanding of ICS prior to Katrina. Senior emergency management officials did receive some ICS training 2-3 days after Katrina made landfall, but this training was too little, too late. It must be done before a disaster strikes.

- There are a couple of inherent conflicts in the execution of the Stafford Act using the architectures of the NRP and ICS. First, ICS is a system where requirements are identified at the lowest level and sourced from the lowest to highest level (county/parish to federal government). This is accomplished by the Incident Commander sending his requirements to the parish/county emergency manager who—in the event he cannot meet the requirements—forwards to the state who—again, in the event it cannot meet the requirements –forwards to FEMA. In disasters this system is often trumped by the political process where a Mayor, or parish president/county executive goes directly to the state looking for resources to meet his/ her requirements. Some of this is understandable and even needed, but only as a matter of exception. In my observation, as matter of rule in disaster operations, the political process is often dominant over the ICS process——resulting in duplicative systems. Second, the Stafford Act is based on the federal government providing supplemental assistance to a state when the state is overwhelmed. ICS is based on who can best meet a need. The ICS is a much faster process however, operations can be slowed somewhat by following the Stafford Act process. For example, the federal government, as a matter of process, fills needs identified by the state. Many times, we know what the needs are but wait for the state to fill these needs through its state processes such as the Emergency Management Assistance Compact. If the state fails to meet these needs they come to FEMA, but it's often several days after the resource is needed that the federal government gets the request. Ultimately, as a result, when FEMA fills the need it then appears and is reported that we have filled the need too late or failed to act in a timely fashion.

Need a trained, staffed, and equipped RESPONSE TEAM.

- **There is insufficient training for personnel involved in response operations.** What we currently have is a very basic "101" type of training. We need more individual and team training institutionalized for response team members. We also need to have advanced training on critical operations such as chemical, biological, radiological scenarios. Katrina taught us that we also need to look at other operations as well, such as evacuation, continuity of government, re-entry, and restoring critical and economically important infrastructure.
- ***The overwhelming majority of federal response team members have response operations as their secondary or tertiary jobs.*** *Response team members must have response operations as their primary jobs—just like first responders such as firefighters and police.* Although we're not first responders, we do work in a time sensitive environment that; requires us to operate at a very fast-paced tempo, requires us to make quick decisions with 10-20% of the information we normally need to make decisions, requires us to have instant credibility and trust in our co-workers because the decisions we make can be and are life saving; requires us to have the requisite skill set to operate and make decisions based on our expertise (there is not enough time to go look something up in a response operation). Anything less than a primary job is inadequate for a response team member.
- **We do not have the force structure needed for major response operations.** Regions do not have sufficient manpower to man a Regional Response Coordinating Center and a full ERT simultaneously. Most significantly, there is a gross lack of senior management and planning expertise in response operations. Critical functions such as operations, logistics, and plans are very thin and one-deep in many cases. With the present force structure we can perform the mission for small and medium size response operations, but in large-to-catastrophic operations we are woefully short. As an example, for Katrina, we did not have sufficient Emergency Response Team staff for a day shift, much less a night shift. During the early days of the disaster, team members had to work around the clock in many cases.

- **We rely on contractors and Department of Defense (DOD) for most all of our equipment needs except for the US&R and DMAT teams.** We need communications equipment for connectivity to state and local communities, a limited number of vehicles for movement of critical commodities, and immediate access to helicopters and busses.

Need to change financial management of disasters. *The current system does not provide the maximum benefit and return on the taxpayers' dollars, hampers states in the execution of their responsibilities, and degrades disaster response & recovery operations.* States don't have enough financial strength to fund disasters up front. They often times have to call a special legislative session or take administrative actions to fund the non-federal share of disaster operations. One of the most common actions states take is to request the federal government pay 100% of disaster costs. **We must never do this; we should give states loans in lieu of 100% federal funding.** When states pay no costs, the result is invariably greater inefficiency, waste, and unwarranted costs. Loans by their very nature would give states greater ownership in disasters, force more fiscal discipline, and eliminate any gamesmanship in the decision making process at state/local levels during disaster operations. These loans should be provided up front.

Simplify the Public Assistance (PA) process: Currently, it takes too long and costs too much to put federal dollars into the hands of the communities that need cash quickly to rebuild and recover. The administrative process for executing PA is fundamentally flawed, resulting in confusion, high administrative costs, and delays in processing applications for federal funding. The process needs to be simplified and modeled after the private sector home insurance program where an inspector visits the site, makes an inspection, determines the federal share, and writes the check all in one visit.

Simplify Individual Assistance (IA) process: The current IA process is too complicated, which often results in confusion and delays in timely delivery of cash and assistance to individuals.

- We confuse citizens who apply for assistance. When we send them a check, we don't send them an explanation of eligible expenditures with the check. The explanation is sent in a subsequent mailing. . . . Also, some of our explanations can be confusing. For example, if someone has insurance, we send them a note saying they are "ineligible". Actually, they may or may not be eligible; what we should be telling them is that they need to settle up with their insurance company before we can provide those funds.
- The sub-caps, within the $26,200 total, should be removed. The repair and replace caps are insufficient in many cases and result in giving disaster victims poor options. For example, if someone had $12,000 in damages to his home, we could only give him approximately $5,000 to repair the home because of a cap. That means the victims don't have enough money to fix the home, so we provide a travel trailer for a temporary residence. . It costs much more, in money and time lost, to put the victims in a travel trailer than it does to give the victims the $12,000 originally needed to repair their home.
- Temporary Housing is not cost effective or customer oriented. It can cost up to $90,000-$100,000 per mobile home for a group site (total costs for site preparation, hauling and installation, and cost of home) and $30,000 -$40,000 for a travel trailer. Families can stay in this temporary housing for up to 18 months, but then they are required to leave. If we gave them the maximum of $26,200, many of these families would have the resources to find permanent housing immediately. This would allow them to quickly get on with rebuilding their lives and afford them an immediate permanent housing solution. It also saves the U.S. taxpayer hundreds of thousands of dollars.

Leadership & management. We need to invest in leadership and management within FEMA. We do not have a system for "growing" leaders and managers. We should do a better job of rotating and cross-training our management and other staff between Headquarters and the field. This would help us better execute our mission and facilitate communication, create better understanding of roles and responsibilities, help us to more effectively achieve all of our objectives, and foster teamwork. This would also help avoid putting policy makers in the awkward position of having to develop, modify, execute, and explain policy without the field experience necessary to understand and appreciate the ramifications of policy decisions.

Emergency management is not a simple system; accordingly, there are no simple solutions. To have an effective national disaster response structure, we must have a viable and fully integrated and coordinated local, state and federal capability. If any of these links in the emergency management chain breaks, the system starts to break down. If we cannot have viability at all three levels of government, then we should change the system.

Thank you for giving me the opportunity to share a field perspective of the response to Hurricane Katrina and of emergency management in general. I will be pleased to answer your questions.

Philip E. Parr

Federal Coordinating Officer
Hurricanes Katrina and Rita

TESTIMONY BEFORE THE SENATE HOMELAND SECURITY AND GOVERNMENTAL AFFAIRS COMMITTEE

Good morning Chairman Collins and members of the committee. My name is Phil Parr and I want to thank you for the opportunity of testifying before you about my experiences and the response to Hurricane Katrina. The views expressed in my testimony are my own and do not necessarily reflect the views of the Department of Homeland Security.

Presently, my position with FEMA is that of Federal Coordinating Officer. I have been involved with response and emergency management for the past 26 plus years. I was sworn in as a member of the New York City Fire Department in 1979 and rose through the ranks to attain the level of Chief Officer in 1999. During my tenure with the FDNY and particularly during my tenure as a Chief Officer I served in many capacities including but not limited to: fire and emergency ground commander, operations, planning for Y2K scenarios, and as a Deputy Director in the NYC Office of Emergency Management. I have played an active role in countless disasters and crisis situations, to include the 9/11 attack at the WTC where I was on scene prior to the towers collapse.

Since January 2004, I have been a member of the Federal Coordinating Officer (FCO) program assigned to FEMA Region 1, New England. In this role I have served in various capacities including FCO for three Presidentially declared disasters.

Before I continue with my testimony I think it important to mention that I've always taken great pride in my years of service as a member of the NYC Fire Department. I could not imagine serving in any other position in which I would serve with the same feeling and pride. However, during my tenure with FEMA, the dedication to service as displayed by its members and their care for disaster victims has allowed me to serve with the same pride and satisfaction that I experienced in my previous 25 years of public service. So it is with that passion that I speak before this committee and I thank you again for the opportunity to do so.

On Saturday August 27th I was informed that I would be the Emergency Response Team Advance Element team leader for the State of Texas. My team was composed of personnel from FEMA, Region 1 (New England), and we were instructed to rendezvous in the Region 6, Regional Response Coordination Center in Denton, TX. Sunday August 28th. Soon it became clear that Texas was not in the path of Hurricane Katrina and that members of my team and I would be assigned as the lead element in New Orleans, La.

I flew into Louisiana, immediately following the hurricane passing Monday 29th of August with a contingent of my team, and Tuesday morning on the 30th we helicoptered into the Superdome. Our mission was three fold: (1) form a unified command with the State (as represented by the Louisiana National Guard), and the City of New Orleans; (2) maintain visibility of commodities ordered; and (3) build out a base from which FEMA teams could be formed to locate and assist in the hardest hit Parishes.

To accomplish these goals we were to meet a Mobile Emergency Operations and Communications Vehicle and use that as a base of operations and communication. Due to extensive flooding in the City our communications vehicle was unable to enter the Dome and this severely hampered our operations. Despite this, and while working under the most difficult of circumstances, we were able to assist the National Guard in maintaining a supply of food and water to Superdome evacuees, all were fed and provided water, and even with limited communications, facilitate the arrival of what was to become over the next four days, a thousand bus convoy to evacuate the City of New Orleans to start the day after our arrival.

The FEMA Disaster Medical Assistance Teams treated hundreds, and identified seriously injured and special needs patients who were evacuated via air and ground assets throughout the operation. In addition, several meetings were held with the Mayor and his staff, ranking National Guard Officers on scene, and other Federal Officials to include DOD and the USCG, this facilitated the initiation of a unified command structure. Due to the enormity of the event, not all of our initial goals were met, and a delay ensued in placing teams into other hard hit parishes, which I believe took place that Friday/Saturday.

I have been asked whether FEMA was overwhelmed and could our response be considered slow. To consider the latter first, I must say in my opinion No. FEMA teams (response, management, medical, and Urban Search and Rescue) were in position in four States pre-land fall. Commodities were staged close to the impacted areas and in some cases the hand-off to the State had already taken place. In addition and as previously mentioned, FEMA mission assigned Emergency Support Function (ESF) 1, the Department of Transportation, and they verified that by 3 September 990 busses were in service performing evacuations and it is estimated that 66,825 persons were transported by that date. The number of buses grew to over 1100 in the next two days. Also, we pushed our logistics capabilities and in the first 6 days after Katrina's landfall, FEMA sent more trucks of supplies for Katrina victims than were delivered in Florida the entire 7 weeks of the response to the four hurricanes last year.

Were we overwhelmed? The simple answer is, Yes. But what needs to be understood is that at any disaster the initial response always feels overwhelmed.

I must draw on my experience as a local responder to give you an example on a small scale of what I mean, and then a larger one. The police officer who pulls up to a two car accident with severe injuries while he operates alone waiting for help is overwhelmed. The fire officer who pulls up to a burning structure with people trapped inside is overwhelmed. But the true professional while responding and operating knows that he is constantly sizing up the situation, gaining intelligence, shifting strategies, modifying plans, and calling for assistance where needed to meet unfulfilled needs whether expected or unexpected.

I would like to refer back to the disaster of 9/11 and its effect on the emergency personnel operating at the WTC. First, it must be remembered that within the 369 square miles of NYC are the resources of a State with a strong central government. There are over 35,000 NYC police officers, about 13,000 firefighters and emergency medical personal, and these numbers only begin to enumerate the assets available to the City. No other city in the country can begin to come close to the responders contained within the City of New York. The response to the attacks on the towers was immediate; the enormity of the task at hand was overwhelming. Then with the collapse of the towers it was chaos. Emergency services within NYC regrouped almost immediately and restarted operations, but a full coordinated plan took days. The WTC complex was thirteen acres. The landfall of Hurricane Katrina affected four States and covered an area of some 90,000 square miles, an area the size of Great Britain and it affected millions of persons. Effectively Louisiana was hit by two disasters, first a devastating hurricane along with its associated blast damage, and second a catastrophic flooding event caused by levee failures. Hurricane Katrina was the most devastating disaster to hit our country. We were all overwhelmed, the City, affected Parishes, the State, and the Federal Government.

What can FEMA, individuals, local government and States do to be more prepared? First, it must be realized that response to any crisis or disaster is the responsibility of every individual and form of government in this country. Emergency Management is more than just coordination. It is about partnership with all entities previously mentioned. Each of us plays a vital part and any one of us who fails in our part fails in that partnership. That failed responsibility must be picked up by one of their partners and that causes delay, confusion and lack of coordination.

For FEMA's part it is my belief we have not done what is needed to get that message across to individuals, locals and States. We've worked to create an image that Uncle Sam will be on your door step with Meals Ready to Eat (MREs), water and ice before the winds subside. We've created an expectation that in a large or no-notice event (such as, a terrorist attack or earthquake) we can never hope to meet. As an agency we must help our partners understand their role in the emergency management cycle (as many States and locals do now). To this end I believe we can do much with conditional and competitive grants to State and local governments to achieve this.

Generally, because response is immediate and local, FEMA's primary role in disaster is recovery. With some notable exceptions, what is described at the Federal level as response is in actuality "response support" (i.e. supplying life saving and life sustaining commodities such as food, water, ice, generation etc.) with local and State responders performing what we traditionally call response but as an agency we can do better in the response role. Primarily, I believe this can be accomplished by a shift in attitude and training by some in management and decision making roles in this agency. I believe since the 2004 hurricane season this was recognized and initiated under our previous Director (FIRST TEAMS and Type 1 or Response FCOs) and I believe this will be ably continued by the now Acting Director Chief David Paulson.

In another area of improvement, FEMA has initiated a total asset visibility system whereby truck loads of commodities can be located via satellite transponder and tracked more closely. This system must be put fully online before our next hurricane season. But more is still needed. While knowing where our trucks are is important, but if they do not have the ability to get to the impacted area with their lifesaving material due to storm damage or hostile environment their value is diminished. Preplanning and providing force protection in coordination with State and local officials must be considered in providing critical deliverables to impacted areas.

We should also recognize that FEMA is a small agency, especially when compared with other Federal agencies, but its strength lies in the fact that the National Response Plan (NRP) identifies it as the coordinating agency for the entire Federal Response. I believe more drills (familiarization, table top, and other) are necessary between FEMA and other Federal Agencies to help clarify roles and responsibilities under the NRP and in their critical ESFs. Understanding their contribution and role in the Emergency Response Team structure is essential for effective response. These crucial elements must be established and become routine to help ensure a better coordinated Federal package can be delivered to the States to assist them in their response.

Additional standardized and practical training must be provided to personnel who may be asked to serve on response teams at the county or local level. Training programs and expectations that build on practical experience from this and previous operations, with input from States, must be provided to FEMA staff who may be needed to assist at the local level in response operations.

As with any operation I hope that as an agency we can make changes based on lessons learned, both positive and negative. I would also hope that State and local officials will review their emergency management procedures and also adopt necessary changes to make their response to disasters more effective. Finally, each citizen has a personal responsibility to be prepared and follow warnings from their local officials.

Thank you for this opportunity to speak on this subject. I would be happy to answer any questions at this time.

William L. Carwile, III

Federal Coordinating Officer,
Hurricane Katrina Response and Initial Recovery Operations: Mississippi

Testimony Before the Senate Committee on Homeland Security and Governmental Affairs
On Preparation for and Response to Hurricane Katrina
on December 8, 2005

Good morning Chairman Collins, Senator Lieberman and distinguished members of the Committee. My name is William Carwile. Thank you for inviting me to testify today on the response and initial recovery operations for the Hurricane Katrina disaster. I served in Mississippi from August 29, 2005 until October 15, 2005; first as the Federal Coordinating Officer (FCO) for the Federal Emergency Management Agency (FEMA) and later as the Deputy Federal Coordinating Officer. I recently retired from federal service am currently affiliated with the Center for Homeland Defense and Security at the Naval Postgraduate School in Monterey, California. I am testifying today as a private citizen.

Hurricane Katrina was the worst disaster to strike the United States in recent history from a response standpoint. While the 9/11 World Trade Center terrorism attack was a terrible disaster, it did not truly test the national response system because of the limited area of damage and the small number of survivors requiring medical attention and mass sheltering. Hurricane Katrina devastated large areas in three states and resulted in massive displacement of the population. The extent of damage from Katrina presented many challenges that had not been encountered in previous disasters and there are many lessons to be learned. I applaud your efforts to gather this information. I hope my testimony today will make a contribution. I will offer my perspective from the field, working in disaster areas with state and territorial counterparts, where I have spent almost all of my nine years with FEMA.

I joined FEMA in October 1996 as Director of the Pacific Area Office in Honolulu, Hawaii and first served as a Federal Coordinating Officer in 1997. I was one of the first members of the Federal Coordinating Officer cadre that was created in 1999 to provide pool of trained professional emergency managers for the federal side of disaster response and recovery. In addition, I was appointed one of the first pre-designated Principal Federal Officials (PFO) by former Homeland Security Secretary Tom Ridge. My recent disaster experience includes: Operations Section Chief for the 9/11 World Trade Center Response; FCO for the 2003 California Wildfires and FCO for each of the four 2004 hurricanes that struck Florida. Prior to my service with FEMA, I served almost three decades as a Special Forces and Infantry officer in the Regular Army including two combat tours in Vietnam. I was a training support brigade commander and the Defense Coordinating Officer for Defense Support to Civil Authorities operations in the Pacific prior to my retirement from the active duty as a colonel. I have provided a list of my military and emergency management experiences.

Background. Prior to the Hurricane Andrew disaster in Florida, FEMA had almost no response capabilities. It was viewed as a recovery agency that basically wrote checks to cover eligible expenses. Even after Hurricane Andrew only a very small portion of the Agency was dedicated to response operations. Those few who were committed to response were sometimes referred to as "cowboys" by some in the agency which was largely staffed for mitigation, recovery, and flood insurance. Since 9 /11 there has been more emphasis on the federal role in disaster response, particularly a response to a terrorism event. In 2005 the National Response Plan was adopted; Homeland Security Presidential Directive - 5: Management of Domestic Incidents, February 28, 2003 strongly advocated the use if the Incident Command System (ICS) for the management of disasters.

Pre-landfall Actions. On August 26, 2005 I was deployed to Mississippi as Hurricane Katrina approached the coast to serve as the Emergency Response Team - Advanced leader and, upon declaration, the Federal Coordinating Officer. Later on August 26th I discussed the situation with the designated Operations Section Chief, Bob Fenton, and we decided he would stop at the FEMA Region IV Headquarters in Atlanta on the way to Jackson to submit Action Request Forms for the response resources that might be required if the storm were to strike Mississippi. Our initial requests were based on the storm track projected by the National Hurricane Center and our years of experiences managing large tropical storm events. In addition, he requested that Meridian Naval Air Station, Mississippi be made available as a staging area for Federal resources.

I arrived in at the State of Mississippi Emergency Operations Center (EOC) on August 27th where I linked up with the Federal Emergency Response Team – Advanced and the State Emergency Response Team, under the leadership of Robert Latham, the Director of the Mississippi Emergency Management Agency (MEMA). The priority at that time was to monitor the progress of evacuations along the coast.

Three Disasters. When a state experiences a large disaster the Governor may request a Presidential Disaster Declaration. One important point to remember about Katrina - there were three separate disasters declared by the President as Hurricane Katrina struck the Gulf Coast - Louisiana, Mississippi and Alabama. Each disaster was different in the amount and type of destruction, the number of people affected, and each State's Constitutional relationships with its local jurisdictions (counties and municipalities). Each disaster had a different person in charge - the Governor of the declared state. As the FCO, I was appointed by the President as his representative to provide support to the Governor. To effectively perform this duty, it is essential to form a close working relationship with the Governor and his or her representatives. In Mississippi this relationship, under the leadership of Governor Barbour, was a strong one, based on a "unified command" approach that worked. The FCO has no authority to direct the state response, but does provide technical assistance, expertise and is authorized by the Stafford Act to commit federal resources in support of the state's response and recovery. If the state has not already entered into contracts with suppliers for commodities, the federal government can contract for those commodities and the state agrees to pay the non-Federal share, if there is one.

Many things went well in the response in Mississippi however there were also problems. The state / federal Unified Command worked well in Mississippi, but this success was obscured by the fact that, initially, requested resources did not arrive into the State quickly enough for distribution to victims. Also, in Mississippi, temporary housing efforts far exceeded any previous effort but this success was obscured by the overwhelming need and the unacceptably long time people remained in shelters. Below I detail some of my thoughts on what went well and what needs rethinking / improvement.

What went well

1. Organizational structure.
In any large operation, coherent command and control is essential. We, the state and federal partners, established a unified command, organized to reduce span of control by using both geographic and functional branches within our joint operations section, and implemented the joint action planning process.

National Response Plan, NIMS, and ICS in Mississippi In my view, the Katrina response in Mississippi was the first time appropriate portions of the National Incident Management System, the National Response Plan, and the Incident Command System were used in a major disaster response. During the summer of 2005, following the distribution of the NRP, many federal and state emergency managers underwent training on the Plan, and in ICS. Fortunately, I and key members of our Emergency Response Team in Mississippi, which was mostly comprised of personnel with whom I have worked for years, had participated in extensive ICS training. Similarly, Mississippi Emergency Management Agency Director Robert Latham and his staff, and most county emergency managers, had recently undergone NIMS and ICS training. One of the key members of our team, Operations Section Chief, Bob Fenton, has long been involved in doctrinal and training development and is truly an expert at how to adapt ICS for large scale operations. Using this training and our experience, Robert Latham and I, and our teams, established a Unified Command beyond that envisioned by the NRP, and began the joint incident action planning process. This established our priorities for each operational period, following ICS concepts. Governor Barbour attended and participated in many of our meetings.

Unified Command – Based on my experience in Florida last year with Governor Bush and his outstanding Director of the Department of Florida Emergency Management, Craig Fugate, I knew the importance of a solid State – Federal relationship through a unified command structure. From the moment I arrived in Jackson, there was a solid State – Federal effort in which Governor Barbour established the strategic goals and direction, and Robert Latham and I and the joint team developed objectives and set priorities for the limited resources. While not completely consistent with the new National Response Plan, which calls for a "multi-entity coordination group", Robert and I believed some entity, in this case a "unified command" should be responsible for setting priorities and making decisions. This approach was further realized through the efforts of our joint Operations Section Chiefs, who fit the resource, State or Federal, to best meet the requirement. Overall, this partnership worked because of the cooperation and complete focus on the mission of assisting victims.

Incident Command System (ICS) Modifications - During the response, we found that some aspects of the NRP did not fit our organizational needs for this joint state/federal response; it was necessary to modify some important aspects of the Plan. It is my belief that "pure" ICS works well for fires and smaller disasters, but some substantial modifications are required for large scale events. Mostly, these revolve around the need for unified command up and down the organization in order to address political and operational realities, and the fact that there may be no local "incident commander" with the capabilities to field a coherent response. ICS, as practiced by the fire service, is based on a "bottoms up" approach in which the Incident Commander and his/her Incident Management Team develop an incident action plan and request required resources. In a catastrophic event this is very difficult if the people at the "bottom" are overwhelmed and unable to fully form coherent response organizations. One modification we made to the basic ICS was to have joint section chiefs in each of the ICS sections - one chief from the state and an equal chief from FEMA; sort of a unified command at the section level. These two individuals worked together to accomplish the goals assigned to their section and reported jointly to the FCO / SCO Unified Command.

We extended this joint command concept to the geographic branch directors and the division supervisors in the local areas. Initially we did not have enough qualified personnel available to provide a federal division supervisor for every county impacted. Pre-landfall we established three divisions along the coast in counties expected to be the most impacted; each division had a state and federal supervisor. Unfortunately, there were insufficient numbers of qualified personnel to place joint division supervisors in each county before landfall. New divisions were added as personnel became available and were trained. The goal was to have coequal federal and state division supervisors in each of the most effected counties.

This was the first time joint division supervisors were co-located with county emergency managers in accordance with ICS doctrine. In Florida last year we did place "County liaisons" in some critical jurisdictions. This was primarily to provide advice to the local officials. County liaisons were just "liaisons" and did not have the authority to direct State or Federal resources. It is my belief that fielding geographic Branch Directors and Division Supervisors worked extremely well. I would recommend that an effort be made to capture the experiences of the individual branch directors and division chiefs involved through interviews, similar to the Army's oral history program, to find out what worked and if changes to the National Response Plan should be made to accommodate the realities of a joint state/federal ICS response.

Emergency Response Team Organization. While there are significant portions of the Incident Command System that do not fit the political and operational realities of a catastrophic disaster, there are ICS concepts that make sense. Among these are: the organizational structure for the team: Command, Plans, Operations, and Logistics. The fifth element, Finance and Administration deal with personnel and funding issues and should not be rolled into a "joint" State – Federal operation.

Span of Control. In ICS doctrine the ratio for span of control is one supervisor to five to seven subordinates. We achieved this through the use of geographic and functional branches. Initially, it was difficult to achieve this ratio because of the lack of available personnel.

2. Temporary Roofing Program. This program involves teams of contract personnel professionally installing high quality plastic sheeting over damaged roofs. This was first used extensively following Hurricane Andrew and again in Hurricane Georges in Puerto Rico, and last year in Florida. While this is seems to be an expensive way to get a blue plastic roof on a home, it enables families to reoccupy their houses until more permanent repairs can be made. The US Army Corps of Engineers was mission assigned to perform this task and achieved results far exceeding any we saw in Florida last year. The Corps, using contractors, was able to install close to the goal of 1,000 temporary roofs a day.

3. Emergency Management Assistance Compact (EMAC). EMAC is a compact between states to provide emergency support across state lines in the event of an overwhelming disaster. The number of personnel that was brought to the State by MEMA under EMAC was almost 25,000. These responders performed close to 900 missions. Many of the personnel who came were first responders, i.e. law enforcement, fire, and medical that were critically needed because of the devastation to the infrastructure of the most affected communities. Note: While EMAC has proven invaluable in recent hurricane events, I must caution against over reliance in a pandemic event or a scenario in which the nation is threatened with multiple terrorist attacks.

4. National Guard The National Guard of Mississippi and the Guardsmen and women from the several other states performed superbly throughout the response. They moved into the most impacted areas quickly and cleared key roads into the area. There efforts a search and rescue were responsible for saving many lives. The entire distribution of commodities mission was theirs. When resources were available, they got them into the hands of victims as quickly as they could. In the area of public safety, they worked well with law enforcement and because of their join efforts, early public safety issues were promptly dealt with and isolated looting ceased after about two days. Major General H. Cross, the State Adjutant General became an invaluable member of the unified command and through it coordinated most of the operations of the Guard.

5. Public Safety/Law Enforcement. Under the leadership of the Commissioner of Public Safety, George Phillips, the members of the public safety community aggressively moved into areas immediately after the storm passed and saved many lives and brought order. This was a very difficult mission as much of the public safety infrastructure, police and sheriffs' stations, patrol cars, and communications had been destroyed in the coastal counties. The men and women of law enforcement, as well as fire fighters and emergency medical personnel did an extraordinary job under incredibly difficult conditions. Commissioner Phillips, because of his many key resources, became an important part of the unified command leadership.

6. The National Hurricane Center's Proactive Approach. Max Mayfield, Director of the National Hurricane Center, should be commended, not only for the accuracy of the Center's forecasts and his media availability to inform the public of the danger of Katrina, but also for his aggressive and uncompromising approach to personally contacting state and national leaders in an effort to begin early evacuations and preparations for what he knew could be a catastrophic event. His actions undoubtedly saved many lives.

7. Others Many individuals and groups contributed to the success of the response and should be integrated into any disaster plan. Volunteer agencies, faith based groups, Navy Seabees, Coast Guard search and rescue and EMAC teams, animal rescue organizations to name a few of the important entities that participated in the response.

What needs improvement

1. Insufficient Disaster Resources. Generally, in discussing resources, there are "teams" and "commodities".

Teams. These include medical (Disaster Medical Assistance Teams, Disaster Mortuary Teams, Veterinary Medical Assistance Teams), Engineer Teams, i.e. Prime Power, and Urban Search and Rescue Teams. These teams were staged on August 29th in Mississippi and were adequate, with the exception of the mortuary teams which were unfortunately unable to provide sufficient support on as timely a basis as required in Hancock County, where there was a shortage of refrigerated facilities and body bags.

Commodities. In Mississippi we were in the unfortunate situation of managing shortages of commodities for the first nine to 10 days. The most critical commodities early on are always: water, food, and ice. We were woefully short of these items during the first critical days of the response. Fortunately, the State of Mississippi had purchased some items for Hurricane Dennis that had not been used. Also, very importantly, the State of Florida had pre-positioned considerable resources to be used in the Panhandle, in anticipation of a hit by Katrina. These commodities were provided by Governor Bush to Governor Barbour under the Emergency Management Assistance Compact and they were able to provide some relief to the victims in the coastal counties. This was very fortunate because, despite requests that were submitted pre-landfall, only about 25% of requested water and ice arrived in Mississippi during the period of September 1st to the 9th. Food, in the form of Meals Ready to Eat (MRE), was also in short supply. I speculate the shortages were the result of an overly centralized logistics system that was overwhelmed by requirements of three large disasters; however, I really don't know what the problem was. I requested that we, in Mississippi, be permitted through our own federal procurement capabilities, purchase commodities throughout the nation to supplement those being provided by the centralized system. I was authorized by FEMA Director Michael Brown to do so.

The factors contributing to the slow delivery of commodities should be examined and addressed for future disasters. Possible solutions reside much better planning efforts between State and Federal emergency management logisticians and operations personnel, the assistance and advice of DOD strategic logistics planners, and much more robust private sector partnerships, e.g. the US Army LOGCAP or USAF AFCAP programs. It is also possible for states to enter into their own contractual agreements with the private sector for procurement and delivery of response commodities. The federal share is reimbursable by FEMA. Florida routinely enters into such agreements.

2. Inadequate Numbers of Qualified Staff.
The inability to field experienced personnel in Mississippi had a major impact on our operations there. FEMA needs many more trained people who can deploy to disasters. Both career professionals and temporary disaster assistance employees (DAE's) performed their jobs well in Mississippi, especially in the first chaotic days after landfall. However, there were not enough trained people to adequately staff all of the positions. Of all the shortfalls that I had to manage as FCO this was the most difficult. This paucity of qualified personnel hurt us in both the response

and recovery phases of operations. In the response, it became apparent that while the deployment to the coastal counties of division supervisors enabled us to maintain linkage with these important jurisdictions, we needed experienced personnel in the cities, e.g. Gulfport, Biloxi, and Pascagoula. There were simply no personnel available in the FEMA system for these duties, to establish Disaster Recovery Centers, or perform traditional outreach activities by Community Relations. We quickly decided to utilize individuals from other Federal agencies, i.e. the US Forest Service and city firefighters from throughout the country, who have trained, operationally oriented personnel. Interestingly, when we used other agency personnel, we often heard that there were no "FEMA" people in a specific community. Despite our use of other agencies, we never had sufficient personnel to meet requirements. Not only were we competing with requirements in Louisiana, there simply are insufficient personnel, trained as members of coherent teams, to meet the requirements of a catastrophic event.

The lack of personnel also meant that many FEMA employees worked long days and weeks without relief because there was no backup. This leads to exhaustion and impacts on morale. In my experience the vast majority of FEMA employees are dedicated to helping victims and communities; they will sacrifice their personal lives to help; however, this will eventually take a toll and many leave the agency for personal reasons. Any organization needs to appreciate and nurture its most important asset - its people. I know that in Florida I had people working longer hours and deployed for many months longer than I would have liked because there was no one to fill their positions. Many of those same people are now in the one of the Gulf States.

The reconstitution of a viable, well-trained Federal disaster workforce, organized into coherent teams, must be a major focus going foreword.

3. Communications.
Despite the deployment of the FEMA Mobile Emergency Support System (MERS) to provide robust satellite communications systems down to the coast to support our forward command post, and hand held satellite phones in county emergency operations centers, our communications capabilities were far short of what was required for effective operations. There should be a concerted effort to provide redundant communications capabilities down to all key responders. Reliance on cell phone and land line technologies proved unworkable. While satellite communications offer quick solutions, there should be a renewed interest in radio systems to provide both long and short range capabilities. Whatever systems are selected, they must be compatible, hardened against man made and natural hazards, and fielded down to local emergency managers in sufficient numbers.

4. Recovery Programs

The Public Assistance program provided under Section 406 of the Stafford Act is far too cumbersome and time consuming in terms of getting funds through the States down to the impacted communities. The program could be totally revamped, borrowing heavily from private insurance company practices. Debris removal – Category A of the Public Assistance Program is, in my experience, one of the most difficult and contentious aspects of disaster recovery. Initially, clearing of roads, waterways, airports, and other areas that are important to the welfare of the community is a priority. Later, the issue of debris on private property arises. In most cases, debris removal on private property, including private enterprises (golf courses, gated communities) has not traditionally been reimbursed under the Stafford Act. In the past we have allowed individuals to move their debris from private to public areas where FEMA reimbursed the federal share for removal. The amount debris in Katrina along the Mississippi coast was so overwhelming - entire neighborhoods pushed into heaps and strewn with potential hazards that we were asked to cover the removal of debris on private property. To accommodate this situation, FEMA Headquarters issued policies to facilitate reimbursement for debris on private property while addressing rights of entry and "hold harmless" legal issues. While we could authorize paying the federal portion for the debris removal under the section of the Stafford act that allows removal of debris on private property for health and safety concerns, reimbursing applicants for the bulldozing of property without an owner's permission was not allowed. The entire issue of Federal reimbursement for debris removal should be addressed in a comprehensive manner. In my view, it will be difficult to get around the "rights of entry" and "hold harmless" portions of the policies but perhaps all debris should be eligible for some specific number of days following a major disaster. In any event, this is an issue requiring clarification.

Individual Assistance programs. These programs should be retooled to become much more "customer" friendly. Simple measures like reviewing all the form letters send to applicants in order to make them understandable and less foreboding. One example is the form letter that informs an applicant that he/she is "ineligible" for FEMA assistance if they have insurance. It is not clear that the applicant may be eligible if they suffer uninsured losses. Another relatively quick fix would be to remove the "caps" on categories of individual assistance put in place by the Disaster Mitigation Act of 2000. The maximum grant amount of $26,200 should remain in place but the entire amount should be able to be applied to repair or replacement costs making the applicant's home habitable. As it is now, only a portion of this amount may be applied to repair costs. This may force a family into seeking temporary housing from FEMA in the form of a travel trailer or mobile home.

5. Temporary Housing. Temporary housing was a major challenge, unlike other programs, this is the primary responsibility of the Federal Government. In Mississippi, we were able to get many more individuals into adequate living conditions more quickly than in any other disaster, now over 24,000 travel trailers and mobile homes, with over 60, 000 individual occupants. However, many people remained in shelters for unacceptably long periods of time.
We tried many approaches to the problem of housing displaced victims, including cruise ships, hotel vouchers, rental assistance and travel trailers. Despite the close cooperation of local officials and hard work by many individuals, people were understandably frustrated by the delays in acquiring adequate housing. Temporary housing is a very complex problem that requires a great deal of planning and resourcing if it is going to be better next time. There will always be delays in a disaster as widespread as Katrina where tens of thousands of people are displaced. I believe that we need to rethink this issue and come up with a suite of options that can be adjusted for the disaster and local conditions. I would suggest that some kind of permanent housing be included as part of the mix in severely devastated areas, but this is not currently authorized under the Stafford Act.

6. Need for well understood operational methodologies. There has been no operational doctrine developed by FEMA in over four years. In my view, there is no clear understanding of the responsibilities of each level (Washington, the Regions, deployed Emergency Response Teams) and how they are to interact. This lack of operational doctrine results in unacceptable levels of overlap, double and triple ordering of resources, and interminably long video teleconferences and conference calls. While some of these are necessary, they can disrupt field operations. In my opinion, well understood and defined operational methodologies based on doctrine would minimize the need for lengthy video teleconferences and would achieve other efficiencies.

There was criticism of the "bureaucratic red tape" in the Katrina response. One thing I learned from the Army - in any operation, particularly a chaotic environment, there needs to be a balance between "going outside the system" and following a plan and a procedure. It is my belief that in any disaster response there needs to be a well disciplined, systematic approach based on a solid plan that is sufficiently flexible for a variety of situations. Experienced personnel know where the pitfalls are and can make decisions where flexibility is required. Doctrine, policies, training, and exercises should be developed that meet the needs of a trained and ready workforce.

7. The role of the Federal Coordinating Officer.
Under the Stafford act, the state is assigned a FCO. The FCO is an integral member of the Unified Command and must be present in the state to co-conduct Unified Command meetings and participate in decision making as problems arise. On September 21st I was notified by David Paulison, acting Director of FEMA that my appointment as Federal Coordinating Officer for Mississippi had been terminated; this action was officially recorded in the Federal Register on that date. Vice Admiral Thad Allen was appointed the FCO for each of the three declared Gulf States - Louisiana, Mississippi and Alabama. I remained in Mississippi as the Deputy Federal Coordinating Officer to Vice Admiral Allen. This may not appear to be a major issue; however, the FCO as the President's representative has specific authorities under the Stafford Act. Without those authorities I was uncomfortable performing the duties of FCO. Vice Admiral Allen was understandably focused on Louisiana and New Orleans leaving Mississippi without the attention of a fulltime, onsite FCO during a critical period in the response. I would recommend that the role of FCO be clarified and that, while the Stafford Act indicates that the President may appoint any individual as FCO for a declared disaster, there should be a pool of pre-designated trained individuals to fill that role. Within the context of ICS these could be "typed" much like US Forest Service Incident Commanders and have standing emergency response teams. Some can be trained and credentialed for smaller disasters and others for catastrophic disasters; some for natural disasters and some for terrorism events.

8. Multi-state jurisdictions - Katrina exposed a weakness in the National Response Plan - there is no specific discussion of multi-state disaster management options. The PFO is the Homeland Security Secretary's representative. The NRP training manual describe the duties of the PFO as to 1) ensure that incident management efforts are maximized through effective and efficient coordination, 2) provide a primary point of contact and situational awareness locally for the Secretary of Homeland Security and 3) provide a channel for media and public communications and an interface with appropriate jurisdictional officials. More specifically, the PFO does not become the Incident Commander or direct or replace the incident command structure. He also does not have directive authority over the Senior Federal Law Enforcement Officer (SFLEO), Federal Coordinating Officer (FCO), or other Federal and State officials.

I have and served as both FCO on disasters and as PFO and Deputy PFO on large exercises and National Special Security Events. In my understanding, the PFO is not an operational position and the PFO has no authority over the implementation of the Stafford Act.

In my opinion, what is really needed in a multi-state operation is not a single entity to direct operations in two or more states; each state should be assigned an independent FCO to work directly with the Governor and his/her representatives. In a multi-state disaster scenario, what is required on the Federal side is an adjudicator of resource conflicts and a source of situational awareness for the National leadership. I believe ICS doctrine may provide a model for this organizational requirement. The Area Command in ICS is an ad hoc organization that requires no operations section, as it does not direct actions, but does have logistics, plans, and finance and administration sections. Its primary role is to provide situational awareness and adjudicate resource shortfalls between and among competing incident commanders. This may be a useful model to be examined to fill this void in the NRP.

The lack of clear doctrine for multi-state events needs to be addressed and a multi-state management plan for the federal response needs to be developed and exercised. This will require the cooperation of emergency management at the state level. In my experience, while state leaders are very supportive of one another during disasters, they would not like to be part of a "Regional approach" that would in any way inhibit direct dialog with the National leadership in Washington. Additionally, most State leaders believe that theirs may be unique circumstances and that the conditions in their state's warrant individual attention and solutions.

Conclusion I hope these thoughts will help the committee with its mission. I had limited time to prepare my remarks and will be glad to provide any additional information or clarification the committee requests. My colleagues and I at the Center for Homeland Defense and Security at the Naval Postgraduate School would be pleased to offer assistance to the committee. The center has both faculty and mid-career professionals doing graduate work who are dedicated to the study of issues I have discussed today. In addition, I am in the process of completing a white paper on the Katrina Response in Mississippi, similar to the white paper I prepared after the 2004 Hurricanes in Florida. These papers deal primarily with technical details internal to FEMA. Both can be made available to the committee.

In closing I would like to comment on questions raised about the competence of FEMA personnel in the wake of Hurricane Katrina. In my career in FEMA I worked with many dedicated and highly competent individuals who were committed to serving both their country and the victims of its disasters. Many routinely give up holidays, anniversaries, birthdays, and normal family lives to help others. I am proud of my service with FEMA and especially proud of the many individuals whose heroic efforts helped the people of Mississippi and other states in which I have served. This is especially true of the "Division Supervisors" (Eric Gentry, Mike Beeman, and Jeff Lusk) we sent down with their State counterparts to coastal counties before landfall, the Urban Search and Rescue teams, the medical teams, and others who joined us down on the coast immediately after the storm passed. Just as with the men and women who serve in our military, these dedicated individuals deserve thanks for what they do.

A disaster can bring out the best in people. There are thousands of stories of individual acts of heroism and kindnesses during Katrina, just as there were when I was in New York City. Mississippians helped their neighbors, hundreds of local officials reported for duty, having just lost everything, and all around the country, volunteers left their lives behind and headed for the Gulf Coast to help. We should realize that the government and those wonderful volunteer agencies can never provide all the aid that is needed in the immediate aftermath of a catastrophic disaster. We all need to better prepare ourselves and our families and be ready to help our neighbors.

I am heartened by the fact that you are taking measures to capture the lessons learned, both good and bad, from Katrina in Mississippi in order to make real changes so that the next time, and there will be a next time, elected and appointed officials will be able to better support the needs of victims.

I thank the committee for undertaking this important work for the nation. I will be glad to try to answer any questions you may have.

Committee on Homeland Security and Governmental Affairs
EXHIBIT #1

Under Secretary EP&R/FEMA Meeting with FCOs

Wednesday, June 30, 2004

Mr. Secretary,

Thank you for taking the time to meet with us and for valuing our input to the management of our core mission.

As you sensed, all is not well in Denmark! While any one of the nineteen issues can pose a challenge by itself, when you connect the dots there are six major areas that come to the fore and present real obstacles to command, control and core mission accomplishment.

The 19 items can generally be grouped into 6 major areas:

1) **Move FCOs & the Declarations Unit into Regional Operations**
2) **Address Operational Training Issues**
3) **Re-Combine the Response and Recovery Divisions**
4) **NIMS ICS Implementation**
5) **Address Comptroller Issues**
6) **Address several Human Capital Issues like Pay Cap Waivers, Comp Time & DAE Travel**

Consolidate Items:

1. Issue: Lack of Engagement & involvement of FCO's in Response Operations.

Discussion: As indicated by your informal poll of Wednesday, 6/30/04, there was a near unanimous belief by the FCO's that you, FEMA, DHS and the Nation would be better served by having them as a direct report to the office of the Under Secretary.

- When FEMA reorganized over a year ago there was debate on where the FCO cadre belonged. Working under Recovery during a **recovery** only disaster proved to be a workable decision. However, when an ERT is deployed to a disaster that has a response component to it, i.e.: Hurricane Isabel, the Response Division feels that they are not represented at the disaster site. In addition, when response issues arise at the disaster, Response feels the need to work directly with the state and circumvent the ERT assigned.

- In the Final draft of the NRP the requirements of consequence management and crisis management are combined, as are the response and recovery parts of a single incident. And the FCO is identified as a critical position in all parts of our all-hazards posture.

- The separation of Response and Recovery into two separate Divisions, with the FCO's placed in Recovery, has caused an increasingly challenging issue for the connectivity of a "single incident" response. This is further articulated in specific examples identified below, and includes being informed of the Under Secretary's visit to a state where a DFO is operational, access to the Under Secretary in urgent operational matters, best use of FCOs while in a non-deployment status, training and lack of preparedness of our National Emergency Response Teams.

- Having the Response Division responsible for FIRST and the ERT-Ns, but not the Declarations Unit or the FCOs, has resulted in the ERT-Ns being identified as number ten (10) of the top ten (10) items for Response and is indicative of their lack of focus on the next "big one". It has also caused the appointment of a non-FCO as a National Team Leader.

- FIRST's operational requirements are spelled out in the NRP and have recently been identified as the number one priority for Response. However, their operational line of control and authority continues to be discussed in a very vague manner. Originally they were to be IMTs and Response had 4 FCO positions for the team leads. **However, contractors now fill the positions and they are called FIRST members, but aren't the positions still inherently governmental?**

    - Certainly the concept of placing a small cell on the ground to meet with the Incident Commander and offer assistance absent Stafford Act authority could create legal and procedural issues as well as unrealistic expectations that can best be vetted through a robust Regional Operations.

- The Declarations Unit and the appointment of an ERT Team Leader/FCO/ICS IC are at the forefront of a Stafford Act Response and/or an NSSE, and yet they are disconnected until Recovery has an opportunity to become engaged.

- When an FCO is assigned a disaster, the Stafford Act, the NRP & NIMS ICS state he or she represents, and is held accountable, by the President and all of FEMA.

Recommendation: Move the FCOs and the Declarations Unit to Regional Operations and select an SES as the manager. By its location as a direct report to the Under Secretary, there would be a robust one-stop-shop for the connectivity of all three elements. It would allow the revitalization of Regional Operations while providing a single point of contact for critical operational alignment and policy dissemination issues as identified above. Further, the communication between Response and Recovery should improve dramatically and the transition from response to a recovery operation will be improved. In keeping with this connectivity, the FCOs should be a direct report of the Regional Director when not deployed, and not relegated to a Division or Branch.

2. Issue: Unpreparedness of National Emergency Response Teams.

Discussion: Before 9/11: ERT-N FCO's were SES', dedicated budget, support staff at HQ, equipment provided for members.

First Year After 9/11: $1.2M dedicated to ERT-N preparedness for the one (ERT-N Red) Team. Provided for development of OPLAN, training, equipment purchase, exercises, and outreach w/Regions.

Now.

- ***Unprepared Teams. Zero funding for training, exercises, or team equipment.*** Three additional Teams formed (Blue, White, and NCR) all being inaccurately reported as "Green". More realistically: White and NCR are Yellow, at best. Red and Blue - Low Green. NCR authorized 74 personnel, 45 rostered. There is a persistent but unconfirmed rumor that NCR will be de-activated.

- ***ERT-N OPLAN is out of date.*** Must be revised to reflect NIMS. Revision should be a priority since not one word of response doctrine, I.e. Manuals, Field Operations Guides, guidance, etc. has been published in over two years.

- ***The ERT-N's are, "the Director's Teams".*** They provide the only practical, expeditious option for the Director to field a cohesive team of his best people to handle the next "big one". While certain new teams are being contemplated, only well trained FEMA led emergency Response Teams can provide solid leadership, down where it matters, in large-scale natural disasters or terrorist attacks. Even FIRST members are folded into the ERT and the JFO, under the FCO once a declaration has been issued and an FCO identified.

Recommendation: Provide resources (four budget proposals submitted over last 18 months with no acknowledgement) in order to rebuild your teams to levels appropriate for them to manage the next "big one".

3. Issue: Re-Establishment of a single R&R Division.

Discussion: Clearly, this alone would not solve all the current problems. However, it would facilitate the re-focusing that is necessary to regain some of the efficiency that has been lost. A robust R&R Division under a single Division Director brings the essential elements of our core mission under one roof. Communications both horizontally and vertically would increase. The disaster experience base of the current staff would be maximized and preserved for the development of future staff. It would also lend itself to a more orderly transition to the NIMS ICS.

Recommendation: When it is prudent to do so, re-establish a single R & R Division.

4. Issue: Need for a single command structure to satisfy our all-hazards response requirements.

Discussion: Over the past two years there have been several attempts by the Response Division to create their own response command structure outside the existing ERT-A/N response elements. Two examples of this are the proposed and dropped Incident Management Teams (IMT) and the current proposal of the FIRST teams. In each example, there was no attempt to incorporate these teams into the existing ERT command structure identified by the NRP to coordinate the Federal response.

Command and control at any disaster will be severely hampered by having parallel or competing command structures with different reporting chains. It is imperative that this not happen.

Recommendation: Identify and publish a concept of operations for any team assignments that may operate within an established ERT structure. Using the NRP, the NIMS doctrine and proven ICS organizational structure, any specialized team should be assigned and report to the Operations Section of the ERT.

5. Issue: Need to establish the NIMS ICS structure with an implementation date.

Discussion: HSPD-5 directs all federal agencies to operate under a single management system. NIMS, based on the fire service's Incident Command System, was chosen as the incident management system on which all federal agencies responding to an all hazard disaster will operate. However, until the Under Secretary publishes an organizational structure and implementation date that FEMA will operate with, every disaster will continue to operate using different organizational structures.

- The decision to mandate the use of this system, while necessary, may not have been thoroughly thought out. The proposed structure is based on the U.S. Forestry Service model. It can be cumbersome and lacks the flexibility to be effectively applied to all-hazards disaster management. The implementation process has not yet considered the inherent cultural environment that is being exposed to these changes. In short, we are taking a work force that is culturally and historically "recovery" minded and force feeding them a system that is not only completely foreign but changes their mindset to a more "response" centered system. This is also absent any significant funding for the necessary training.

Recommendation: Send out an all hands e-mail providing an organizational structure that is NIMS ICS compliant, and establish an aggressive implementation date and training schedule that will allow FEMA regions and ERT deployments to gain valuable experience prior to the height of the hurricane season.

6. Issue: Lack of training, policy guidance and standards for the Operation Section of the Emergency Response Team (ERT).

Discussion: Under NIMS ICS, the Operations Section is responsible for providing a rapid and effective Federal presence during the early stages of a catastrophic disaster or terrorist/WMD incident. The Response Division coordinates with the affected Region and State on the immediate execution of disaster response missions. The Emergency Response Team (ERT) directs the delivery of Federal assistance programs and services, coordinating all operations in support of the response to the emergency.

Recommendation: FEMA HQ should have a dedicated unit within Response's Operations Branch that is responsible for; (1) writing and promulgating regulations, policies, and standard operating procedures, (2) being a resource for the Operations Section Chiefs in the field, (3) providing the Operations Section Chief staff person to the EST, (4) working with the training folks to ensure regional staffs are adequately trained and the development of an establish a credentialed system, which incorporates Task Books for rostered personnel to affirm there ability to perform specific tasks required for there rostered function, (5) coordinating with OGC, OFM, and the IG to ensure the Mission Assignment process is fiscally responsible and accountable, (6) ensuring effective accountability at all levels and within individual functional areas during incident operations, (7) coordinating with the Regional Operations, and (8) numerous other related tasks necessary to ensure the processes for response operations are consistent and effective throughout the agency.

7. Issue: Comptroller obtrusiveness in operational areas harms mission accomplishment.

Discussion: The Comptroller is the principal advisor to the FCO in matters of the disbursement and expenditure of federal funds from a financial management perspective. However, during the recent past, and in an institutional manner, HQ directorates have allowed Comptrollers to engage in operational decision making outside of their expertise. The result has been a lack of operational effectiveness and undue conservatism in decision-making, leading to a decrease in operational effectiveness and even paralysis. For example, arbitrary comptroller decisions have resulted in vital management training not being given at a DFO.

- **Example** – Extracted from a recent email – "Bad news - we (EMI) may not be able to support resident courses such as Disaster Field Operations Management (DFOM), Managing FEMA Staff, SitStat, etc. Bonnie Butler, Disaster Operations Recovery Branch of EMI has been approached by the Office of Financial Management (OFM) about this type of training being conducted at DFOs. Bonnie is my direct supervisor who reports to the EMI superintendent, Steve Sharro (Ultimate decision makers on training decisions and actions).

  I would advise the FCO, Lou Botta, check with the comptroller to see if there are some issues regarding the expenditure of disaster dollars for training activities of this type. Evidently some of the comptrollers have raised some concern to OFM who in turn raised these same concerns/issues to EMI. I know we had a very successful trainings conducted at the DFO with DFOM and other courses. However, the comptrollers have been concerned about this for some time and now it has raised the hackles of some folks" etc.

Recommendation: Remind Comptrollers that their role is to advice and make recommendations to the FCO. Under NIMS ICS, as a general rule, a comptroller does not have his/her own individual authority.

8. Issue: Are support functions driving operations?

Discussion: The thrust of actions/policies/procedures from the headquarters should be aimed at streamlining processes and, where possible, making the job easier for our people in the field. Numerous examples abound where HQ (in many cases DHS HQ) actions compound problems rather than resolve them, i.e.,

- Problems associated with blocks on FEMA purchase cards delay effective operations.

    - **Example** - On Guam for Lupit, we could not readily buy 1) pharmaceuticals needed on Yap, 2) containers necessary for storage at Guam warehouse, 3) wood chippers needed for debris removal, etc. Examples at Des Moines, for DR-1518, included the rental of forklift needed to unload trucks during DFO setup and the rental of a motel room at Mason City for Public Assistance PAC site, etc.

- Requirement to use U.S Bank C.A.R.E. system to validate transactions made with FEMA purchase cards easily doubles amount of time contracting/purchasing personnel at DFO have to spend on bookkeeping.

- DHS restrictions on water in the DFO are unduly severe. The current rule, as it appears in the draft purchase cardholder's manual, is that if there is ANY potable water source in the DFO, this is sufficient. It further states that it doesn't matter if the water is discolored, foul tasting, or malodorous. Some DFOs have only one water source, and it may be far-removed from some of the work areas. Controls on individual bottles of water are not the issue – employee welfare can be well served, at a modest cost, by renting coolers from local vendors with five gallon bottles of water for common use. Ironically, the draft rule specifically authorizes the installation of additional water fountains, but not the purchase of containerized water.

Recommendation: Make sure that these and other items are not only "captured" through the RAMP process but that they are quickly addressed. The revitalization of Regional Operations, and the consolidation of the FCOs and the Declarations unit into it, would provide a robust single point of contact on the Undersecretary's immediate staff; this would better support and resolve operational issues.

9. Issue: Pay Cap Waivers.

Discussion: Through the summer of 2003, pay cap waivers were essentially automatic from HQ, as they should be. At least the first two weeks of ANY disaster operation requires 12-16 hour days while we set up the DFO, DRC's and get to know the issues of our communities and the state(s).

There is also the issue of parity with other federal agencies. General observation shows that the folks from the Corps of Engineers get a pay cap waiver for the duration of any FEMA deployment and we should get nothing less.

Recommendation: Every disaster declaration should come with a pay cap waiver for, as a minimum, the amounts of time necessary to take it up through the first FULL pay period. This would be a minimum of two weeks and could be up to nearly four weeks. There should also be a procedure to allow for the pay cap waiver to be extended for those operations that are particularly difficult.

10. Issue: Compensatory time off.

Discussion: FCOs deploy on short notice for an unknown duration to locations that are often Spartan. The need to take care of life's other interruptions, such as family, social obligations, medical procedures, financial transactions, etc. continue and often require us to be 'home' (not everything can be accomplished over the internet). There is also the need to simply rest and relax...to decompress after an arduous deployment.

As it currently stands, DAEs have a dubious advantage in that they can make themselves unavailable for a period of time after a disaster operation. Full-time employees cannot. We are forced to use personal leave if we want time off even though, in the fulfillment of our employment requirements, we usually leave our families and communities for several months at a time. Even with the 18 hours our supervisor is allowed to give as an "award", the leave accumulation schedule is not a generous one. This could be mitigated through the accumulation of Comp Time while on a disaster deployment. The Armed Forces figured this out decades ago and allows up to two weeks administrative leave following an extended deployment. We should have a similar benefit that, as is the case with so many other low cost initiatives, would improve morale and increase retention.

Recommendation: Develop a Compensatory Time policy and incorporate appropriate controls to ensure it is not abused. Such guidelines could include a maximum time off per deployment and establishing a window of time to use the comp time.

Note: HRD is presently drafting a policy document.

11. Issue: Benefits for DAEs.

Discussion: DAEs continue to ask about health insurance benefits and other "equal treatment" items like holiday pay and pay for travel home from a disaster. If we could offer some benefits we should be able to recruit, train, and **retain** qualified staff.

Recommendation: Direct HRD and OGC to finish, in a timely manner, their current reviews of these items for SAEs. Once complete, direct HRD to develop the appropriate policy memos.

12. Issue: Fencing of DAE's.

Discussion: It appears that the fencing of DAE's is more prevalent now than before ADD was centralized. The main reason given for centralizing ADD was to discontinue fencing of personnel and make all DAEs available for assignments wherever they are needed. Since HQ has been calling the shots, fencing has occurred mainly for "what if" situations and has penalized DAEs who would have been available for deployments.

Recommendation: Through Regional Operations and the Regional Directors, place a stronger emphasis on making DAEs available. Monitor through our weekly availability tracking system.

13. Issue: Logistics/IT support for Field Operations.

Discussion: Expectations for iPass greatly exceed reality. Lots of problems with iPass in the field:

- Dial-up rate through hotel switchboard averages 19kbs in Iowa -- far too slow for our Public Assistance people to work in NEMIS.
- New users, or those users whose passwords have expired, have to be re-registered every time they go out to a disaster -- process can take up to eight hours through the National Help Desk.
- All laptops have to be configured with iPass at the DFO -- time-consuming process that slows down DFO and DRC establishment.
- All users need iPass training before they go to the field -- conducted on an ad hoc basis by IT people who have to be pulled from other duties (DFO & DRC setup).

- "Cheat sheet" hip-pocket instructions for users run from four to twelve pages in length -- too much for users with rudimentary computer skills to absorb and retain.

- If feasible, use existing hotel/motel high-speed service for iPass access. Must support multiple access VPN users.

Recommendation: Have HQ IT address user agreements with hotels to allow the use of existing high-speed Internet connections for our field users. Identify **one** primary point of contact for all iPass issues at HQ. The DFOs need "one belly button to push" at HQ IT.

14. Issue: Defacto requirement to establish DRC's concurrent with DFO opening strains resources of Log/IT.

Discussions: Expectations of our customers dictate that we now stand-up DRCs concurrent with the DFOs. To accomplish this we have cut corners, i.e., using Safety or Security Officer instead of Log/It to perform some of our DRC site surveys, having people work/drive sixteen hour days surveying/setting up DRC's, PAC sites, etc. may be over-taxing and non-prudent.

- MERS has not been formally tasked to support this requirement...they need to be. Denton MERS typically supports this effort, however, not all MERS detachments do. Regions are not manned with sufficient Log/IT personnel to support this.

- Actions associated with iPass further exacerbate problem, i.e., we now have to register iPass users, configure laptops for iPass, establish computer lab for training, conduct iPass training for users, etc. Again, all concurrent with DFO/DRC setup.

Recommendation: Have MERS plan for, expect and manage to this new standard.

15. Issue: "Trip-wire" to call the Under Secretary directly from a DFO.

Discussion: In the course of operations, items arise that may demand direct contact between the FCO and the Under Secretary. At present, there are many barriers to such contact. There are levels of bureaucracy between the FCO and the Under Secretary that tend to keep issues from reaching his level. It is not suggested that every issue be elevated, but FCOs should have greater latitude to call the Under Secretary without fear of reprisal from others within the hierarchy.

Recommendation: Direct communications be authorized between the under Secretary and the FCO during DFO operations. This should not be seen as an anomaly but rather a normal course of business when warranted. (Note: It is not suggested that the Under

Secretary be inundated with calls from DFO's, but the FCO's having the freedom to call when warranted.) Certainly under the Regional Operations concept there would be a direct link and connectivity within the Office of the Under Secretary.

16. Issue: FCO's not being informed of the Under Secretary's visit to a state where we have an operational DFO.

Discussion: On occasion the Under Secretary has visited states with active DFOs and yet the FCO had not been informed of the visit. The lack of sharing of this information has lead to the FCO being seemingly uniformed when questioned by the SCO, the Governor's office and the congressional delegation.

Recommendation: Whenever the Under Secretary is scheduled to visit a state where an active DFO is present the FCO should be informed. Even if the FCO is NOT to be a part of the official visiting party, at least by making the FCO aware he or she can talk intelligently about the visit when asked. In addition, the FCO may have valuable information for the Under Secretary. There would be greater connectivity, and less chance of this sort of issue arising, with the FCO's being placed in Regional Operations.

17. Issue: Appropriate threat assessment information would facilitate FCO readiness in case of a major terrorist incident during field operations.

Discussion: DFO security structure is not equipped to deal with present realities. There is little or no way for FCOs to obtain intelligence assessments that could directly affect field operational readiness as well as the DFO's ability to reconstitute to answer to a terrorist incident.

Recommendation:

- Install secure telephones in DFOs.
- Provide Security Officers in DFOs with the appropriate security clearances to liaison with law enforcement and intelligence communities.
- When appropriate, provide classified threat assessments to FCOs during the conduct of field operations.

18. Issue: Provide FCOs with commensurate authority to deal with unforeseen administrative and personnel circumstances during field operations that presently create undue delay and affect mission accomplishment.

Discussion: The FCO should have the authority, responsibility and accountability to make administrative decisions impacting field operational effectiveness and safety. Too often these decisions are now relegated to headquarters. For example, the 50-mile waiver for TDY reimbursement should be delegated to the FCO. The FCOs accept the responsibility and accountability that comes with the position. Again, they should have the authority to make those decisions that, in their opinion, are in the best interest of the their operation and the safety of their personnel.

Recommendation: Seek to remove bureaucratic obstacles where delegation of authority to FCOs may help the success, effectiveness and safety of disaster field operations.

19. Issue: Dual Lodging.

Discussion: We are occasionally confronted with situations where it is necessary to get authorization to pay for lodging in two locations simultaneously. These generally arise when someone has to travel across state or out of state on short notice, for a short duration (one or two days) on official business and there is not enough time to pack-up and move out of his or her current lodging facility. At other times, it is known that this is going to happen well in advance but it is akin to cruel and unusual punishment to have someone pack-up and leave their long-term lodging for only a day or so.

By allowing occasional dual lodging, the government can save a considerable amount of money. In those disasters that cover a wide geographic area, and damage is widespread, it is more cost effective to use a DAE from a central location to travel to the far reaches of the state to take care of business and then return. This saves us the costs associated with deploying one or more additional people to service these areas.

It has been noted that HQ is extremely reluctant to approve these requests even though the travel associated with the request should always be in the best interest of the government, otherwise it should not be occurring.

Recommendation: It is recommended that FCO's be authorized to approve dual-lodging waivers. Appropriate controls can be devised to ensure this is not abused such as: cannot exceed two or three nights lodging, must be certified by memo signed by the FCO and included with the travel reimbursement documentation, etc. This is a very low cost item that could improve the efficiency of operations.

End

Committee on Homeland Security and Governmental Affairs
EXHIBIT #8

U.S. Department of Homeland Security Federal Emergency Management Agency **Action Request**	See Reverse for Paperwork Disclosure Notice	OMB No. 1660-0047 Expires November 30, 2007

I. Who is Requesting Assistance? (Completed by Requestor) EOC DR 316

: LT Gary Marchese NOPD/LA — Temporary Phone/FAX #:

Permanent Phone: [redacted] — FAX# [redacted]

Requestor Organization: New Orleans Police Department — E-mail:

II. Requested Assistance ☐ See Attached

NOPD requests the following supplies
400 x M4 weapons, 25,000x 5.56mm ball ammunition, 300 x Type II A anti ballistic vests (various sizes)
1500 x pairs of black military boots (various sizes) 1000 x hand cuffs, 100 x laptop computers
2000 x New uniforms BDU Dark blue types of various sizes 200 x Crown Victoria Police Cruisers (Vehicles)
Also request approx 35,000 gal of fuel per week be delivered to NOPD gas station at 4800 General Meyer, West Bank

Quantity: See list — Priority: ☐ 1 Lifesaving ☐ 2 Life sustaining X 3 High ☐ 4 Medium ☐ 5 Normal — Date/Time Needed: 09/07/05

Delivery Location: NOPD, 1500 Perdido, New Orleans LA 70119
POC: LT Gary Marchese NOPD/LA 504-658-8700

Site POC: LT Gary Marches or Warren Riley — 24-hour Phone #: [redacted] — 24-hour FAX # [redacted]

State Approving Official signature: *Robert [signature]* — Date: 5 Sep 05

III. Sourcing the Request – Review/Coordination (Operations Section Only)

☒ Ops Review by: *Wayne [signature]*
☐ Log Review by:
☐ Other Coordination by:
☐ Other Coordination by:
☐ Other Coordination by:
Immediate Action Required: ☐ Yes ☐ No
Date/Time Assigned:

☐ Donations ☐ Procurement
☐ Other (explain) ☐ Interagency Agreement
☐ Requisitions ☐ Mission Assignment

Action request assigned to: ☐ ESF#: ☐ Other:

IV: Statement of Work (Operations Section Only)

OFA Action Officer: — 24-hour Phone #: — 24-hour FAX #:
FEMA Project Officer: — 24-hour Phone #: — 24-hour FAX #:

Justification/Statement of Work: NOPD requests listed supplies, in order to continue mission of re-establishing law and order in New Orleans.80% of NOPD officers are working with bare minimum equipment and uniforms. Many lost their personal uniforms during the flood. This equipment is essential to reestablishing the NOPD and maintaining the law and order of New Orleans.
This request was communicated from the NO City Mayor through COL de Graff Deputy Chief of Staff G3/5/7 1st Army on 5 SEP 05
See attached email correspondence

Estimated Completion Date: — Cost Estimate:

V: Action Taken (Operations Section Only)

☐ Accepted ☒ Rejected — ☐ Accountable Property Coordinated with APO
Disposition
APPLY FOR REIMBURSEMENT THRU P.A. *FAXED TO LA EOC*

VI. Tracking Information (FEMA Use Only)

NEMIS Task ID:
Action Request #: — Received by (Name & Organization):
Program Code/Event #: — State: — Date/Time Submitted: 9/6/05 1:15p — ☐ Originated as Verbal

Instructed to submit. If time would permit…we would normally do lots and lots of research…I can't believe that DOJ does not have some program that looks at this…

Your Friend and probably enemy

Wayne

Close

From: de Graff, Christian, COL, DCS, G-3/5/7 [SMTP:christian.degraff@[redacted]

To: Yingling, John A., MG, CMDGRP

Cc: '475dce@[redacted]

Subject: Re: Request for Support, NOPD

Sent: 9/5/2005 5:18 PM **Importance:** Normal

Sir,

Got reject on his email, but got through on the DCE.

V/r

Chris

Christian de Graff
COL, GS
First Army
Deputy Chief of Staff
G3/5/7
[redacted]

-----Original Message-----
From: Yingling, John A., MG, CMDGRP <john.yingling@[redacted]>
To: de Graff, Christian, COL, DCS, G-3/5/7
<christian.degraff@[redacted]>; Hickey, James R., COL, CMDGRP
<James.Hickey@[redacted]>
CC: 'Damon.Penn@[redacted]' <Damon.Penn@[redacted]>;
'anthony.daskevich@[redacted]' <anthony.daskevich@[redacted]>
Sent: Mon Sep 05 18:12:41 2005
Subject: RE: Request for Support, NOPD

Tony,
Are you working this??? Any push back?

MGY

-----Original Message-----
From: de Graff, Christian, COL, DCS, G-3/5/7
Sent: Monday, September 05, 2005 5:52 PM
To: Hickey, James R., COL, CMDGRP; Yingling, John A., MG, CMDGRP
Cc: 'Damon.Penn@[redacted]'; 'anthony.daskevich@[redacted]
Subject: Request for Support, NOPD

Just met with mayor. His Police Department has the following needs:

400 M4s

25K 5.56 Ball
300 Vests
1500 Military Boots of various sizs
Fuel delivered to gas station at 4800 General Meyer on the West Bank.
1000 Hand Cuffs
100 laptop computers
2000 New Uniforms, Dark Blue type, various sizes
200 Crowne Victoria Police crusers.

Need to work in collaberation with FEMA.

CG said, Give me the list and we'll get it.

Chris

Christian de Graff
COL, GS
First Army
Deputy Chief of Staff
G3/5/7
[redacted]

09/06/2005 13:59 FAX 225 925 7501 LHLSEP 001

****
*** ERROR TX REPORT ***
****

TX FUNCTION WAS NOT COMPLETED

TX/RX NO	4313
RECIPIENT ADDRESS	89408985231
DESTINATION ID	
ST. TIME	09/06 13:58
TIME USE	00'00
PAGES SENT	0
RESULT	NG #0018 BUSY/NO SIGNAL

State of Louisiana

MILITARY DEPARTMENT

FFICE OF HOMELAND SECURITY & EMERGENCY PREPAREDNESS

KATHLEEN BABINEAUX BLANCO
GOVERNOR

PERRY J. SMITH
COLONEL (R)
ACTING DEPUTY DIRECTOR
EMERGENCY PREPAREDNESS

JADWIN V. MAYEAUX
COLONEL (R)
DEPUTY DIRECTOR
HOMELAND SECURITY

9-6-05
DATE

FAX COVER SHEET

Action Tracking

FROM: Mike Goldsworthy

SECTION: FEMA Ops MA

PHONE: 225- [redacted]

31 FAX: 225- [redacted]

___ No. of pages (including cover page)

ENDENCE BLVD - BATON ROUGE, LOUISIANA 70806
EPHONE: (225) 925-7500 FAX (225) 925-7501

09/06/2005 14:31 FAX 225 925 7501 LHLSEP 001

****
*** TX REPORT ***
****

TRANSMISSION OK

TX/RX NO	4320
RECIPIENT ADDRESS	89408985231
DESTINATION ID	
ST. TIME	09/06 14:29
TIME USE	02'05
PAGES SENT	9
RESULT	OK

State of Louisiana

MILITARY DEPARTMENT

FFICE OF HOMELAND SECURITY & EMERGENCY PREPAREDNESS

KATHLEEN BABINEAUX BLANCO
GOVERNOR

PERRY J. SMITH
COLONEL (R)
ACTING DEPUTY DIRECTOR
EMERGENCY PREPAREDNESS

JADWIN V. MAYEAUX
COLONEL (R)
DEPUTY DIRECTOR
HOMELAND SECURITY

9-6-05
DATE

FAX COVER SHEET

Action Tracking

31

FROM: Mike Goldsworthy

SECTION: FEMA Ops MA

PHONE: 225- [redacted]

FAX: 225- [redacted]

No. of pages (including cover page)

ENDENCE BLVD - BATON ROUGE, LOUISIANA 70806
EPHONE: (225) 925-7500 FAX (225) 925-7501

U.S. Department of Homeland Security

FEMA

Facsimile Transmission

To: LA State EOC Fax: [redacted]

From: Ops JFO Phone: [redacted]

Date: 9-22-05

Number of pages including cover:

ACTION REQUEST FORM

OMB No. 1660-0047
Expries November 30, 2007

I. Who is Requesting Assistance? (Completed by Requestor) EOC-DR-[illegible] 354

Requestor Name/Title/State: Orleans Parish
Temporary Phone/Fax #:

Permanent Phone: [redacted]
FAX #:

Requestor Organization:
E-mail:

II. Requested Assistance (Completed by Requestor) ☐ See Attached

Description of Assistance Requested:
Gas powered golf carts needed to transport firefighters around the staging areas at Zephyr Field.

Quantity: 10

Priority: ☐ 1 Lifesaving ☐ 2 Life sustaining ☒ 3 High ☐ 4 Medium ☐ 5 Normal

Date/Time Needed:

Delivery Site Location: Zephyr Field
6000 Airline
Metairie, LA 70003

Site POC: Steve Jellie
24 Hour Phone: [redacted]
FAX #

State Approving Official signature: [signature]
Date: 9/6/05

III. Sourcing the Request - Review/Coordination (Operations Section Only)

☒ OPS Review by: Phillip Bowen
☐ Log Review by:
☐ Other Coordination by:
☐ Other Coordination by:
☐ Other Coordination by:

☐ Donations ☐ Procurement
☐ Other (explain) ☐ Interagency Agreement
☐ Requisitions ☐ Mission Assignment

Immediate Action Required: ☐ Yes ☐ No
Date/Time Assigned:

Action request assigned to: ☐ ESF #: ☐ Other:

IV: Statement of Work (Operations Section Only)

OFA Action Officer: 24 hour Phone: FAX#
FEMA Project Officer: 24 hour Phone: FAX#

Justification / Statement of Work:

Committee on Homeland Security and Governmental Affairs
EXHIBIT #9

Estimated Completion Date:
Cost Estimate:

V. Action Taken (Operations Section Only)

☐ Accepted ☐ Rejected

Disposition:

☐ Accountable Property Coordinated with APO

TRACKING INFORMATION (FEMA USE ONLY)

eCAPS/NEMIS Task ID:
Action Request #
Received by (Name and Organization):
Program Code/Event #:
State:
Date/Time Submitted:
☐ Originated as verbal

FEMA Form 90-136, NOV 04

Ⓔ Resource Request

Priority: Red-High
*Status: Teal-Pending Operations
*Request Number: Res-160905-570-090605

WHO IS MAKING THE REQUEST?

Requesting Organization: Orleans
***Requestor's Contact Info:** Phone: [redacted]
Fax:
Pager:
Frequency:
Talk Group:
Cell:
Other:
Call Sign:
Related Event/ Incident/Activity: E - Hurricane Katrina

Display All resource requests related to: E - Hurricane Katrina

Display

WHAT IS BEING REQUESTED?

***Resource Type/Kind:** GAS POWERED GOLF CARTS
***Quantity:** 10
Qty Unit Of Measure: each
When Needed:
Mission: NEEDED TO TRANSPORT FIREFIGHTERS AROUND THE STAGING AREAS AT ZEPHYR FIELD.
Resource must come with:
Other:

DELIVER THIS RESOURCE TO:

Site Name
ZEPHYR FIELD
Street Address
City
NEW ORLEANS

United States
Intersection - Street 1
County

Site Type

Apt or Lot No.
State
LA
Zip

Intersection - Street 2
Geographic Area
(Region, District, Campus, etc).

Additional Location Information

GEO LOCATION & MAPPING

Show on Map?
No
Geo Located By
Latitude
Longitude

Contact on scene: (Name, Phone, etc.) STEVE JELLIE
[redacted]
Special Instructions:

FORWARD REQUEST TO:

Individual:
Organization/Location:
Position:

Agency: U.S. Federal Emergency Management Agency
Vendor: Not Related to a Vendor
Summary of actions taken:
Estimated Resource Cost:

DISTRIBUTION

(Unless you limit distribution as specified below, this document is viewable by all E Team users.)

Group:
Individual:

NOTIFICATION

Send Notification? No
Message
Individuals

Groups

Other Email addresses

DATA SHARING

Comment:

Display Data Sharing History

OVERLAYS:

Create Overlay

Available Overlay(s): No Overlays on File

ATTACHMENTS:

Supporting File(s):

No documents found

Web Pages:

Request History:

09/06/2005 at 16:15 CDT - Terry Vallarautto at LOHSEP as Logger 1 changed status to Teal-Pending Operations

Document Created by: LALogger1 on 09/06/2005 at 16:15 CDT Last Modified by: laasstops on 09/06/2005 at 18:35 CDT

State of Louisiana

MILITARY DEPARTMENT

OFFICE OF HOMELAND SECURITY & EMERGENCY PREPAREDNESS

BENNETT C. LANDRENEAU
MAJOR GENERAL
THE ADJUTANT GENERAL DIRECTOR

8 Sept 05
DATE

FAX COVER SHEET

TO: Tom Joyce (Action Tracking)
Kent Baxter

PHONE:

FAX: [redacted]

FROM: Lauren McLane

SECTION: FEMA Ops MA

PHONE: 225- [redacted]

FAX: 225- [redacted]

COMMENTS: Attached are 15 ARFs - Call 703-66[redacted] you are missing pieces

7667 INDEPENDENCE BLVD - BATON ROUGE, LOUISIANA 7
TELEPHONE: (225) 925-7500 FAX (225) 925-7501

****
*** TX REPORT ***
****

TRANSMISSION OK

TX/RX NO	3432
RECIPIENT ADDRESS	89408985231
DESTINATION ID	
ST. TIME	09/08 00:45
TIME USE	06'40
PAGES SENT	27
RESULT	OK

Committee on Homeland Security and Governmental Affairs
EXHIBIT F

TUES 9/30

- STEVE Delrodo
  LOS to TAC (504) [redacted]
- · 800 #1 Pelican
- GARY MOORE (IOC) NRCC (cell)
  #1 Small ECAT to SAB
  RICHARD WEBER
- CO LOCATE W/ Delrodo — HANDLE
  · MEDICAL MISSION   OAM 1
  IYM 1
  - patient evac
  - supplies IN
- TONY ROBINSON OPS CHIEF
- MICH HOLT - NDMS
  COYLE

NRCC - 202 [redacted]

KOL SANZA

NRD OCT
[redacted]
John Gonzalez
Jerso
[redacted]
helso

DHS-FEMA-0084-0000005

- COL KEITH LACASTE
  LOUISIANA FISH & WILDLIFE
- SAMS CLUB PARKING LOT
  AIRLINE HWY N 29° 58.26.5
  W 90° 10'.14.6

Peter Smalley
~~0800~~ 800 [redacted] - [redacted] SAT PHONE

FIBER 40 BRICKS WHAT IS FIBER

* LT CLESI (225) [redacted] - [redacted]
  DOMMIE

* Pump Water out of Boiler room
* BATON Rouge Exhibit Center - Baton Rouge
~ BARGE - Smith & Co.

NDMS

Ron Martin Ops Officer

~~[illegible]~~ Ops Officer on West unit

2 DMAT UNIT - 35 Assigned 1 here

3 STRIKE TEAMS - 5 assigned 1 DMAT 2 Strike 2 DMAT/[illegible] 1 ST

~~[illegible] TRIAGE~~

Seen & Treated - 0140

. Staff has been doing med EVAL

. 18 TOTAL EVAC . 12 in last hr

. 10 - NEED to EVAL not critical

. 20 - NURSING HMES (same N.H.)

. 400 SPEC NEEDS 32 at a time

truck or Boat hasn't started yet

- RUNNING OUT OF SUPPLIES

~~, MED cash~~ &/

. Patient TREATMENT & PHARM CASH

- ON SITE 1 TEAM ASSIGNED

BRIEF SCOTT/BILL

- shelters
  · scaling down
- here 3 days respond
  3 more days
  · people in water

· NRCC

- EOV @ JFO

*NDMS BPT
*(USCG) SAR [illegible]
*general veillon [illegible]

SAMS CLUB

CURT [illegible] [illegible] [illegible] - CRCC

ar's
- command & CTRL
- forward base of operations
  · convention ctr
- EOV on [illegible] Com suite
- Unified command
  [illegible] comm. ctr
- MIA BCP - 2 fix comms

DHS-FEMA-0084-0000008

EOC
WED 8/31

* AVIATION COMMAND & CONTROL *
* LOUD SPEAKERS MEGA PHONES
* 3 SHELTERS
  - South East LA. UNIV
  - LA STATE UNIV. BR
  - UNIV of Louisiana Layfette
  - McNeese UNIV
  - Nichols State
* LOTO Belle Chase Naval Air
* New Orleans Intl AIRPORT
  - shopping ctr parking lots
* computers & personnal to document victims or video camera for ID

* WATER ON ~~[illegible]~~ BASE TRUCK
PEOPLE TO INTERSTATE
* BOATS BRCC tell us
* COMM SUITE ASAP
ERT-A OPPS & LOG

Committee on Homeland Security and Governmental Affairs

EXHIBIT G

Resource Request

Priority: Green-Low
***Status:** Red-Action Required
***Request Number:** Res-145649-960-101005

WHO IS MAKING THE REQUEST?

Requesting Organization: Orleans
***Requestor's Contact Info:** Phone: 504-289-4295 504-658-8700 Chief Matthews
Fax:
Pager:
Frequency:
Talk Group:
Cell:
Other:
Call Sign:
Related Event/ Incident/Activity: E - Hurricane Katrina

Display All resource requests related to: E - Hurricane Katrina

Display

WHAT IS BEING REQUESTED?

***Resource Type/Kind:** Miniature Bus
***Quantity:** 1
Qty Unit Of Measure: each
When Needed: 10/11/2005 at 13:00 CDT
Mission: Transportation for Mayor and approximately 20 Employees to travel to Shreveport on City business regarding evacuees.
Resource must come with: Fuel, Meals, Operator(s), Lodging
Other:

DELIVER THIS RESOURCE TO:

Site Name
Hyatt Hotel
Street Address
Poydras
City
New Orleans

United States
Intersection - Street 1
Poydras
County

Site Type
Emergency Operations Center
Apt or Lot No.

State
LA

Zip
70112

Intersection - Street 2
Loyola
Geographic Area
(Region, District, Campus, etc).

Additional Location Information

GEO LOCATION & MAPPING

Show on Map?	Geo Located By	Latitude	Longitude
No			

Contact on scene: (Name, Phone, etc.) Lou Reese [redacted]
FEMA Scott Ericson ([redacted]

Special Instructions: This request has been Clarified and is being resubmitted for approval.

FEMA has denied this request for support as" Not within the best Financial interest of the Federal or State Government" per FEMA Ops Wayne Fairley

---

Return Thursday 10/13/05 afternoon. Hotel accommodations provided for Driver. State Assistance requested. Resource not available locally. Resource Request approved by Chief Matthews. Input LGM

FORWARD REQUEST TO:

Individual:
Organization/Location:
Position:
Agency: U.S. Federal Emergency Management Agency
Vendor: Not Related to a Vendor
Summary of actions taken: 10/10/2005 at 17:18 CDT LAFEMA Joint Force Operations
Rejected per FEMA Operations - Not within the best financial interest of the Federal or State government.

This request has been Clarified and is being resubmitted for approval.

Estimated Resource Cost:

DISTRIBUTION

(Unless you limit distribution as specified below, this document is viewable by all E Team users.)

Group:
Individual:

NOTIFICATION

Send Notification? No
Message
Request rejected. Please see summary of actions taken.
Individuals

Groups
OHSEP Operations Group
OHSEP Division Chief
Operations Assistant
Other Email addresses

DATA SHARING

Comment:

Display Data Sharing History

OVERLAYS:

Create Overlay

Available Overlay(s): No Overlays on File

ATTACHMENTS:

Supporting File(s):

No documents found

Web Pages:

Request History:

10/10/2005 at 15:11 CDT - Orleans Logistics1 at LOEP as Logistics1 changed status to Red-Action Required
10/10/2005 at 18:32 CDT - Assistant Operations Officer at LOEP as Assistant Operations Officer changed status to Blue-Closed
10/11/2005 at 09:18 CDT - Assistant Operations Officer at LOEP as Assistant Operations Officer changed status to Red-Action Required

Document Created by: LAOrleansLogistics1 on 10/10/2005 at 15:11 CDT Last Modified by: laasstops on 10/11/2005 at 09:24 CDT

Committee on Homeland Security and Governmental Affairs
EXHIBIT H

Resource Request

Priority: Red-High
***Status:** Yellow-On-Scene
***Request Number:** Res-101428-292-091005

WHO IS MAKING THE REQUEST?

Requesting Organization: Orleans
***Requestor's Contact Info:** Phone: 504-658-8700
Pager:
Frequency:
Talk Group:
Cell:
Other:
Call Sign:
Related Event/ Incident/Activity: E - Hurricane Katrina

Display All resource requests related to: E - Hurricane Katrina

Display

WHAT IS BEING REQUESTED?

***Resource Type/Kind:** Air conditioner
***Quantity:** 1
Qty Unit Of Measure: each
When Needed: 09/10/2005 at 10:16 CDT
Mission: RNA - Need portable A/C units to cool approximately 5000 sq. ft. of office space.
Resource must come with:
Other:

DELIVER THIS RESOURCE TO:

Site Name
New Orleans City Hall

Site Type

Street Address
1300 Perdido St # 1W23

Apt or Lot No.

City
New Orleans

State
LA

Zip

United States

Intersection - Street 1

Intersection - Street 2

County

Geographic Area
(Region, District, Campus, etc).

Additional Location Information

GEO LOCATION & MAPPING

Show on Map?	Geo Located By	Latitude	Longitude
No			

Contact on scene: (Name, Phone, etc.) Mayor Ray Nagin [redacted]
Special Instructions: Working issue with Nor E Base Camp Equipment.
1-800-686-5277 POC: Mary Point.

FORWARD REQUEST TO:

Individual:
Organization/Location:
Position:

Agency: U.S. Federal Emergency Management Agency
Vendor: Not Related to a Vendor
Summary of actions taken: Being worked through NOR E Base Camp Equipment 1-800-686-5277. Completed per GSA
Estimated Resource Cost:

DISTRIBUTION

(Unless you limit distribution as specified below, this document is viewable by all E Team users.)

Group:
Individual:

NOTIFICATION

Send Notification? No
Message
Individuals

Groups

Other Email addresses

DATA SHARING

Comment:

Display Data Sharing History

OVERLAYS:

Create Overlay

Available Overlay(s): No Overlays on File

ATTACHMENTS:

Supporting File(s):

No documents found

Web Pages:

Request History:

09/10/2005 at 10:20 CDT - Nicole Roper at LOEP as Assistant Operations Officer changed status to Teal-Pending Operations
09/27/2005 at 17:07 CDT - Melvin Smith at LOEP as Tracker1 changed status to Yellow-On-Scene

Document Created by: LALogger1 on 09/10/2005 at 10:20 CDT Last Modified by: LATracker1 on 09/27/2005 at 17:07 CDT

Committee on Homeland Security and Governmental Affairs
EXHIBIT I

Resource Request

Priority: Red-High
***Status:** Gray-Cancelled
***Request Number:** Res-175051-957-101405

WHO IS MAKING THE REQUEST?

Requesting Organization: LA Dept. of Social Services
***Requestor's Contact Info:** Phone: 225-925-7373
Fax:
Pager:
Frequency:
Talk Group:
Cell:
Other:
Call Sign:
Related Event/ Incident/Activity: E - Hurricane Rita
Display All resource requests related to: E - Hurricane Rita
Display

WHAT IS BEING REQUESTED?

***Resource Type/Kind:** Taxi - Transportation
***Quantity:** 1
Qty Unit Of Measure: each
When Needed: 10/14/2005 at 17:54 CDT
Mission: Provide transportation for 1 person from Summit Hospital to General population shelter, Lifting Up This Temple shelter, 1746 Scenic Hwy, Baton Rouge La. Carolyn Landry 356-3242(shelter contact)
Resource must come with: Operator(s)
Other:

DELIVER THIS RESOURCE TO:

Site Name
Summit Hospital
Street Address
I-12 and O'Neal Lane
City
Baton Rouge

United States
Intersection - Street 1
County

Additional Location Information

Site Type
Hospital
Apt or Lot No.

State
LA
Zip

Intersection - Street 2
Geographic Area
(Region, District, Campus, etc).

GEO LOCATION & MAPPING

Show on Map?	Geo Located By	Latitude	Longitude
No			

Contact on scene: (Name, Phone, etc.) Natalie [redacted]
Special Instructions: Returned to DSS for local action as per Phone conversation with Jim Ballow
----
Deliver 1 person from Hospital to Shelter

FORWARD REQUEST TO:

Individual:
Organization/Location:
Position:
Agency: U.S. Federal Emergency Management Agency
Vendor: Not Related to a Vendor
Summary of actions taken:
Estimated Resource Cost:

DISTRIBUTION

(Unless you limit distribution as specified below, this document is viewable by all E Team users.)

Group:
Individual:

NOTIFICATION

Send Notification? No
Message
Individuals

Groups

Other Email addresses

DATA SHARING

Comment:

Display Data Sharing History

OVERLAYS:

Create Overlay

Available Overlay(s): No Overlays on File

ATTACHMENTS:

Supporting File(s):

No documents found

Web Pages:

Request History:

10/14/2005 at 18:01 CDT - Claire Hymel at LA Dept. of Social Services as EOC Liaison changed status to Red-Action Required
10/15/2005 at 09:22 CDT - Assistant Operations Officer at LOEP as Assistant Operations Officer changed status to Gray-Cancelled

Document Created by: LADSS on 10/14/2005 at 18:01 CDT Last Modified by: laasstops on 10/15/2005 at 09:23 CDT

APEX
TP
9780160772337
76901953 – 80